Thomas Digby Pigott

London Birds and London Insects and other Sketches

Thomas Digby Pigott

London Birds and London Insects and other Sketches

ISBN/EAN: 9783743311213

Manufactured in Europe, USA, Canada, Australia, Japa

Cover: Foto ©berggeist007 / pixelio.de

Manufactured and distributed by brebook publishing software (www.brebook.com)

Thomas Digby Pigott

London Birds and London Insects and other Sketches

London Birds
and London Insects

(REVISED EDITION)

AND OTHER SKETCHES

By T. DIGBY PIGOTT, C B

Naturam expellas furcâ, tamen usque recurret
HORACE. EP. LIB: I.X

LONDON:
M. PORTER, 18, PRINCES STREET, CAVENDISH SQUARE,
1892.

PASSERI DULCISSIMO
QUI,
NEMORUM PRATORUMQUE OBLITUS
PUELLULARUM QUATTUOR
M.D.P., E.J.D.P., H.M.D.P., W.D.P.
DELICIÆ,
AMANS ET AMATUS,
ANNOS TRES IN URBE VIXIT,
HÆC DEDICO.

The greater part of the contents of the following chapters, which have no pretence to be anything more than casual notices of Birds and Insects seen at different times in London and elsewhere, strung together as a holiday amusement, have already been published,—the chapter on "London Birds" more than once.

Having been more kindly received than they deserved, thanks only to the charm which anything to do with Nature has for most of us, especially perhaps, if met with unexpectedly, the notes are reprinted with a few more pages as unscientific as the rest.

The papers on "The Birds of the Outer Farnes," "The Shetlands in the Birds'-Nesting Season," "The Last English Home of the Bearded Tit," and "The Summer Tenants of Dutch Water Meadows," are, with the permission of the proprietors, reprinted from the "Contemporary Review."

T. D. P.

The Mo,
    Sheringham,
        September, 1892

CONTENTS.

CHAPTER I.
	PAGE
LONDON BIRDS	1

CHAPTER II.
THE BIRDS OF THE OUTER FARNES	31

CHAPTER III.
THE SHETLANDS IN THE BIRDS'-NESTING SEASON	49

CHAPTER IV.
THE LAST ENGLISH HOME OF THE BEARDED TIT	77

CHAPTER V.
ST. KILDA FROM WITHOUT	89

CHAPTER VI.
IN DUTCH WATER MEADOWS	109

CHAPTER VII.
LONDON INSECTS	123

APPENDICES ... i-vii

London Birds

" 'Tis always morning somewhere, and above
The awakening continents, from shore to shore,
Somewhere the birds are singing evermore."—*Longfellow*.

WE are so accustomed to associate birds—"the smiles of creation"—with all that is wild and fresh, and pleasant, and unlike a great town, that to speak of the birds of London sounds rather like talking nonsense. It is, however, one great advantage which an ornithologist has over most other lovers of natural history, that there are few places in which he cannot find something in his own particular line to interest him, unless it is in countries where Robins and Tomtits have been too long marketable delicacies, and where, as in some parts of the Continent, woods and plantations are dying off in consequence—lands smitten with worms for having slaughtered the innocents. Longfellow's simile is much too good to be given up merely because, as commentators tell us, the King who died on his throne, as he made a speech to the people, was not the Herod who killed the babes of Bethlehem and added

Childermas Day to the Calendar, but a nephew and namesake only.

London is no exception to the general rule. Indeed, in some respects we are unusually favoured. To begin with, there are, of course, the splendid collections, dead and alive, in the British Museum and Zoological Gardens. There are the bird-stuffers' windows, into which a good proportion of the curious birds shot in the kingdom are sure to find their way. There are Leadenhall Market and the game-dealers' shops, with constantly changing supplies all through the year; and, in hard weather, there are the wild-fowl hawkers about the streets, with great bunches of Stints, Curlews, and Oyster-catchers, doing duty as Snipe and Woodcock, and Pochards and Mallard, and Mergansers, "ancient and fish-like" enough to be smelt across the street, with their tell-tale saw beaks broken to make them Widgeon.

But leaving these out of the question, there are the genuine wild birds of London; and it may, perhaps, be a surprise to some readers to learn that a note-book of those seen by one person in the course of not more than a year in the immediate neighbourhood of Hyde Park, all sufficiently near to be identified without difficulty, included more than twenty species—representatives of five of the six great natural orders into which birds are divided.

The exception was the class of the "birds of prey," the Raptores—"low-foreheaded tyrants"—the first in scientific arrangement, but, according to a modern writer, the lowest almost of all in everything but brute force, because they can neither build nor sing.

Wild birds of prey are not very common in London, but, though it is not every one who is fortunate enough to see them, members of both branches of the family, night fliers and day fliers, are occasionally to be seen.

There was a report in the summer of 1891 that an Eagle had been seen circling over the East-end of London. The rumour was unconfirmed, but was the more likely to have been true, as about the same time a big bird of the kind, which, from the description given, was probably a young white-tailed Eagle, was seen, heading northwards, by two gentlemen fishing in the Stour, at Chartham, a few miles above Canterbury.

Dr. Edward Hamilton, who published in the *Zoologist* in 1879 a carefully compiled paper on "The Birds of London, Past and Present, Resident and Casual,"* numbering in all nearly a hundred, says that "in 1859 a Kite was observed flying over Piccadilly not above one hundred yards high," and mentions as "casual visitors" the Peregrine Falcon, Kestrel, and Sparrow Hawk.

A large Owl—a grave and reverend representative of the night fliers—was apprehended by the police-constable on duty a year or two ago in the Repository of the Public Record Office, and after inquiry discharged with a caution—unfortunately before the species had been determined.

Another, of a species also undetermined, has since been recorded in the newspapers as having taken up its quarters for some days in a tree in the grounds of Guy's Hospital.

We hear in these days much of the struggle for existence which is going on everywhere in Nature, and of adaptations in the forms of animals to the conditions under which they have to carry on the

* A list of the birds noticed in London, based on Dr. Hamilton's paper is printed as an appendix.

fight. There is not a clearer or more beautiful instance of the kind than the wing of a common Brown Owl.

The bird has to hunt close for its prey in the dark. If it cut the air with the noisy flight of a Partridge or Wood Pigeon it would soon starve, for every one of the timid little creatures which are its natural food would take good care to keep out of sight till the danger was past; and so—as an oar is muffled to deaden the splash—Providence has hung a soft, loose fringe of down to the front of the quill. This makes the Owl's flight, as every one who has watched a Barn Owl "mousing" knows, perfectly silent.

Of the second, the "Passerine"—the enormous order into which are jumbled all which cannot be classed as birds of prey or poultry, and which, as a rule, neither climb, nor wade, nor swim—we have a very respectable party constantly in London. Not less than seventeen or eighteen appear in the list of birds seen within the year referred to at the beginning of the chapter, and this does not of course nearly exhaust the number of common visitors.

First come the Thrushes—the most timid, perhaps, of all; but, by one of the apparent contradictions with which all classifications abound, nearly related to the Shrikes, which are the connecting link between the passeres and the birds of prey, and, in their own degree, scarcely less tyrants than the Eagles themselves. Song-thrushes are fairly common in Kensington Gardens and St. James's Park, where they nest regularly, and sing beautifully at times; though, as a rule, they are very shy. During the middle of the day they manage, to a great extent, to keep out of sight; and it is not often, when many people are about, that they show themselves in any considerable numbers. But when the gates are first open, and the early morning dew is on the grass, one may see them, four or five at a time—stamping to start the worms, then hopping for a yard or two, and standing still to listen, with their heads on one side, and their bright eyes sparkling with attention.

Blackbirds, too, are common, though less so than Thrushes, and also nest in St. James's Park. They are, probably, migratory with us, as they are more plentiful at some times of the year than at others.

Fieldfares and Redwings are to be seen occasionally in cold weather; but we have no great supply of berries to attract them, and their visits are short.

But though, by right of their voices no less than the notch in their beaks, the Thrushes claim the place of honour; easily, first among London birds, by numbers as well as impudence, are the Sparrows. Poking about in every gutter, and dusting themselves almost under the horses' feet with all the amusing self-possession of street urchins, they take care not to be overlooked.

But for one quiet house in a corner, the Zoological gardens might be the happy hunting ground of good Sparrows. Dainties are to be had for the stealing all over the place, and even the lions and bears and Eagles are too sleepy and well-fed to resent any amount of petty larcenies. It is a melancholy thing, though, to see the end when it does come. The snakes are fed one afternoon in the week, and five or six tailless Sparrows are a dainty meal. Unlike the rabbits and guinea-pigs, who will nibble and sniff at a python's nose, they seem too wide-awake to doubt their fate for a moment, and crouch together in a corner, the picture of dejection—till, if the snakes are hungry, there is a sudden flutter, and the miserable party scuttle over to another corner, one short in numbers; and one may see a little bunch of feathers, at all sorts of impossible angles, peeping out from a coil of scales. The stroke is almost quicker than the eye can follow.

London Sparrows evidently look upon Corinthian capitals as designed for their especial convenience in the nesting season; and Bishop Stanley tells of one pair which had the impertinence to build in the mouth of the lion on Northumberland House, long ago departed to the limbo of forgotten landmarks where Copenhagen and the "big" Duke on the Arch have since joined him. When the Duke's statue was taken down for removal to Aldershot, in 1884, it was found that more than one bird—like Gavroche in the plaster elephant of the Place de la Bastille—had set up house inside. There was a Sparrow's nest with a newly hatched young bird, and several eggs in the right arm; and, in the elbow of the left, a nestful of young Starlings almost fledged.

The front door of both establishments was a hole in one of the hands.

With all its ragged untidiness, few things are grander in suggestion than a Sparrow's nest on Westminster Abbey or St. Paul's. It carries one back to the days when the author of the Eighty-fourth Psalm watched the birds building in the niches of Solomon's temple—or, more probably looked back on with the eye of memory only from exile by the waters of Babylon—and wrote, in words which have still all the freshness of three thousand years ago, "The Sparrow hath found an house, and the Swallow a nest for herself, where she may lay her young; even thine altars, O Lord of Hosts, my King, and my God." The commonest Sparrows in the Holy Land — *Passer Syriacus* — though not actually the same, are almost identical with our own house Sparrows.

In spite of overshadowing soot, there is a considerable variety to be noticed in the plumage of London Sparrows. One spotted with white and another of unusually light tint, very much the colour of a dormouse, have for more than a year escaped the cats at the foot of the steps by the Duke of York's column. Another, a cock bird, with a tail of almost pure white, had, in the spring of 1892, his headquarters just inside the rails of the Green Park, near Devonshire House.

Of the Buntings, the only two which figured in the lists of birds seen within the year were a cock Yellowhammer, picked up dead in the Green Park, apparently starved to death; and another seen in St. James's Square. . The latter was unluckily very tame, and paid dearly for a meal in the gutter, only just managing to flutter on to Lord Derby's house, much the worse for a cut from a cabman's whip.

In March and April, 1890, the "ill-betiding croak" of the Raven was a familiar sound to West-end Londoners; a fine fellow, who, judged by this tameness and by the fact that several wing feathers were missing, was probably an escaped captive, having for some weeks settled in Kensington Gardens where Carrion Crows are fairly common and not afraid to make free with the Ducks' eggs.

Starlings build in numbers in the hollow trees; and, with a few grey-headed Jackdaws, and poor ill-used Rooks, make themselves generally at home among the sheep, and are as talkative and merry as in the reed beds on the Norfolk broads.

Two very interesting papers on "The Rooks and Rookeries of London," the one by Dr. Hamilton, the other by Professor Alfred Newton, are to be found in the *Zoologist*. The tale told is a sad one, and the conclusion drawn seems only too probable: "The Rooks and Rookeries so pleasant to old Londoners are gradually diminishing and disappearing, and the London Rook to our grandchildren will be a bird of the past."

The story of the Kensington rookery is a sample of what is going on all through London. "In 1836," writes Dr. Hamilton, "this rookery extended from the Broad Walk near the Palace to the Serpentine, where it commences in the gardens, and there must have been nearly one hundred nests." "They are now," he adds, writing in June, 1878, "alas! reduced to thirty-one nests and confined to a few of the upper trees skirting the Broad Walk near the North Gate."

Since then almost every tree in the garden which had a nest in it has been cut down, and until the spring of 1892, when there were encouraging signs of a return, and one pair built again in a tree in the south-west corner, there seemed too much reason to fear that Kensington Gardens had lost for ever one of its greatest interests, and that the colony at Gray's Inn was destined to be the only considerable survival of the great rookeries once common in the middle of London.

A Hawfinch, one of the comparatively rare birds which have apparently of late years become more common, was picked up in St. James's Park on the 28th January, 1890. It was a hen in good condition

About the same time the following year, and in the same place, a Mountain Finch was found, a visitor from the North, uncommon in most parts of England, excepting during unusually severe winters. The bird, a male, was in fine plumage, and, judging by the brightness of its fawn colours and whites, could not have been long in London. A breastbone which felt through the feathers like the back of a knife, told the common tale of starvation.

A pair of Chaffinches were to be seen more than once in April, a year or two ago, very busy collecting moss for a nest, between Victoria Gate and the fountains; and two rather dingy little Blue-Tits were about the same time carefully investigating the trees close by, evidently with the same views. Cole-Tits, too, occasionally show themselves in the Gardens. Both the Cole-Tits and Blue-Tits, the latter in considerable quantities, have been caught in London by Doctor Albert Günther, who, before the removal of the collections from Bloomsbury, occasionally relieved his severer studies of Natural History by setting traps in the grounds of the British Museum. He has also caught Greenfinches and Redpoles.

House Martens in plenty, and with them Swallows and more rarely Swifts and the little brown Sand Martens, play on the ornamental waters. The House Martens build in several parts of London. There were three nests—the marks were still to be seen a year ago—in St. James's Street, two of them over Boss's, the gunmaker's shop, two more in Porchester Place, and three on a blank wall in Upper Seymour Street. "Where they most breed and haunt, the air is delicate," and their mud-houses are a compliment to our improved drainage.

Everyone knows that it is unlucky to disturb a Swallow's nest, but the reason why may not be so generally known.

Old women in Norfolk say that when the birds gather in thousands, as they do in many places before they leave us for the south, and sit in long rows on the church roof, they settle who shall die before they come again. Any one who has offended them during the summer may expect to have his name at least brought forward then for consideration.

Wheatears are occasionally to be seen. Two small parties settled in London for a few weeks in August a year or two ago—one in Hyde Park, the other in the Regent's Park. They are very inquisitive little fellows; and, though they will whisk off their pretty white tails before one gets very near them, they cannot go far without stopping for another good stare. They are trapped in numbers in parts of England and France, in little holes cut in the turf, and commanded by common brick-falls. No bait is required, as they cannot resist the temptation to hop in to explore, and their next appearance in public is probably in vine leaves.

The Kingfisher is, perhaps, the last bird one would expect to see in London. Two have been caught at different times in the grounds of the Museum facing Great Russell Street, and a pair, not long ago, made themselves at home for some time near the ponds in Regent's Park. Others have been noticed more than once of late years on the St. James's Park water.

The country round London is a favourite haunt of Nightingales. They have been known very lately to breed in Battersea Park, and every now and then one finds his way into the one or other of the more central parks. They are curiously capricious in their choice of localities for settlement. It is a real Irish grievance —though no attempt has yet been made to redress it by Act of Parliament—that though to all appearance the country in parts, with its green copses and soft climate, is just what should suit them, they are never heard in Ireland. The cock Nightingales usually land in England eight or ten days before the hens. They sing their best when in expectation only of the happiness to follow them, and are said to be valuable as cage singing-birds only if caught as bachelors. It is touching to hear, on the authority of bird-catchers, who know what they are talking about, that a cock caught after he has paired is useless, and will probably mope till he dies.

A Nightingale in splendid voice gave a series of early recitals a year or two ago in Kensington Gardens. In the heat of the day, when there were too many perambulators about, he kept out of sight, "in shadiest covert hid," but before breakfast sang without any attempt at concealment morning after morning on the same almond tree, not very far from the Prince Consort's Memorial.

His song, poor fellow, was all the more impassioned, because the lady he sang for in all probability existed only in his dreams; or, if they had really met and engaged themselves in the warm winter among the olive groves of the South, thought herself absolved from the engagement, and free to console herself with a less audacious mate in a quieter home beside some Kentish lane, when she heard her lover could wish her to follow him to shameless London.

Alphonse Karr, in his "Voyage autour de mon

Jardin," complains of the misrepresentations which have resulted from slavish imitation of the classics by modern writers. Why, he asks, because poets writing in softer climates spoke truly enough of May as "the month of roses," should every French poet think it necessary to do the same, forgetting that what is true in Greece or Italy is not necessarily true in France, and that, as a matter of fact, roses do not blossom there in any very great profusion before June?

Nightingales have even more just cause to protest. Philomela—as all of us know who are not too far removed from school days to remember anything of our Ovids—was a Greek girl compromised in an affair of a marriage with a "deceased wife's sister." Her position, which was trying enough from the first became unbearable when it was found out that the first wife was not really "deceased" at all, but only put out of sight by the husband, who had cut out her tongue. Both sisters had cause enough for complaint, and because poor Philomela did complain and was changed in pity into a Nightingale, and the poets sang her sorrows; therefore the Nightingale must be sad, and always posing as a love-lorn maiden.

" The melancholy Philomel,
Who perched all night alone in shady grove
Tunes her soft voice to sad complaining love,
Making her life one great harmonious woe."

Milton, the Londoner, steeped as he was in the classics, as a matter of course follows suit, and for him the Nightingale is necessarily—

" Most musical, most melancholy."

But even Shakespeare, of whom we might have hoped better things, could not altogethe free himself; and once, in his writings—though certainly only to put the word into the mouth of Valentine, the lovesick Gentleman of Verona—we find the inevitable

> "Nightingale's complaining note."

By-the-bye, if nothing to do with Milton, and Shakespeare had come down to us but their poetry, we should not have had any great difficulty in arriving at a fairly true idea of the sort of lives the two men lived by merely comparing the manner in which each refers to birds.

Take, for instance—and there are plenty of other passages at least as much to the point—such little touches, fresh from Nature, as—

> "Far from her nest the Lapwing cries 'away,'
> My heart prays for him, though my tongue do curse,"

in the "Comedy of Errors." Or in "Much Ado about Nothing"—

> "Look, where Beatrice like a Lapwing runs
> Close to the ground to hear our conference."

Or—

> "Like an eagle in a dove-cote, I
> Fluttered your Volscians in Corioli."

One hears the clatter of the wings, as the startled Pigeons break out all round.

Contrast these with any of Milton's allusions to birds. "Birds of Jove" driving before them "birds of gayest plume,"—"ravenous fowls" hurrying to a field of battle, or—

> "Vultures on Imaus bred
> Disfledging from a region scarce of prey
> To gorge the flesh of lambs," &c.

It is not necessary to multiply instances. In almost

any page of the writings of either that one opens, the contrast forces itself into notice. The magic wand is the same, but the hands that hold it are very different. Shakespeare touches us, and we crouch with him and hear the Night-jar rattle and the Shrew Mice whistle in the fern in the deer park as we hold our breath to listen for the keepers ; or we stroll along the track of old Aikman Street, across the unenclosed commons of Buckinghamshire, and take Plovers' eggs with a rollicking and not over-respectable company of players on the tramp from Stratford and London ; or loll in the shade and listen to the birds and bees overhead in the branches of the oak trees of Grendon Wood. It is fresh Nature everywhere.

Milton takes the wand and the country changes to the town. We smell the leather of dusty piles of learned volumes, and stand half afraid in the presence of the man who could see in the gloom to report for a Parliament of Devils, and look without flinching at—

> "The living throne the sapphire blaze,
> Where angels tremble as they gaze."

But never, even before his blindness, could have had an eye for a bird.

But to return to the Nightingale's song. It is a libel to call it sad. As a matter of fact, it's the exact reverse. There are in it, of course, none of the blood-stirring notes of war and crime to be heard in the cry of the Eagle, nor does it, like the wail of the seabird on the hungry shore, carry with it suggestions of Robinson Crusoe adventure ; but it is peaceful, self-sufficing, and perfectly happy—home affections and domestic joy set to music. Perhaps to some of us, with boys to start in life, even the curious croak, almost like a frog's, which a Nightingale gives every now and then when the young birds are leaving the nest, but only then, may not altogether destroy the truth of the rendering.

The beginning of the singing of the Nightingale was, in old Persian calendars, the date for the festival in honour of the return of warm weather.

Another night-singing warbler to be found at times in London—a pair were seen not long ago in the meadow between the powder magazine and ranger's house in Hyde Park—is the Sedge warbler, a pretty little bird, not unlike a Nightingale, with a white line above the eye. "The Sedgebird," writes old Gilbert White, "sings most part of the night: its notes are hurrying, but not unpleasing, and imitative of several birds, as the Sparrow, Swallow, and Skylark. When it happens to be silent in the night, by throwing a stone or clod into the bushes where it sits, you immediately set it a singing; or in other words, though it slumbers sometimes, yet as soon as it is awakened it resumes its song." Another pair lately settled for some little time beside the water in Regent's Park. For six or seven years there has been a Fly Catcher's nest in Rotten Row.

The Wren, which, as his name, Regulus, the little king, denotes, has been from earliest times a bird of consideration, is fairly common with us.

It was a Wren who shared with Prometheus the honour of bringing fire from heaven, and more than once since, the family has distinguished itself by taking an independent line in public affairs. In the religious disturbances of Charles II.'s time, the Wrens were on the side of the Protestants, and once, "by dancing and pecking on the drums as the enemy approached," saved the lives of a party who would otherwise certainly have been surprised sleeping, and cut to pieces "by the Popish Irish."

"For this reason," says Aubrey, who tells the tale in his Miscellanies, "the wild Irish mortally hate these birds, to this day calling them Devil's Servants, and killing them whenever they catch them."

The sympathy of the Lapwings seems to have been as strongly on the other side. They were for the High Church party, and made themselves hated for generations in the lowlands of Scotland as much as were the Wrens in Ireland, by disturbing the devotions of the Covenanters, and meanly betraying them again and again to the Duke of York and Commissioner Middleton's men, by shrieking on every possible occasion over the lonely meeting-places on the hillsides.

The golden-crested Wren has occasionally, but not often, been seen on the peninsula in St. James's Park.

A few Robins, and a Lark, seen on two consecutive days in Hyde Park and the Green Park, complete the list of the Sparrow family in the year's notes on which this chapter is based, though there is no doubt that, with a little longer observation, a great many others might have been added. A Night-jar attended one of the evening performances of Buffalo Bill, hawking about for some time near the seats of the other spectators, at the end of May, 1892.

The "climbers" are not well represented, the only one that was noted during the year being a single vagabond Cuckoo, who found his way into Hyde Park on the 8th of May, and left in the direction of Park Lane. Painful as it is to say unkind things of those we cannot help liking and wish to respect, it is unhappily quite impossible to deny that the Cuckoo is out of all measure a disreputable bird

She begins life by murdering her foster brothers and sisters, whom she is pretty sure to shuffle one by one on to her hollow back and pitch out of the nest before she is ready to leave it herself. She grows up—the charge is proved beyond all question—to think as lightly of marriage vows as she does of a mother's duties. Excepting that they have two toes in front and two behind—the distinguishing feature of the class—the Cuckoos have little or nothing in common with the "climbers" proper; but a visit from one of them would be just enough to give us a claim to the Woodpeckers as a London family, even if none of the true Woodpeckers were ever to be seen. Stories are still occasionally told of the little spotted Woodpecker, and more frequently of the commoner Green Woodpecker, having been seen in Kensington Gardens. Probably, at one time, both may not have been uncommon there, but their visits are now comparatively rare. A Green Woodpecker was heard and seen in Hyde Park in November, 1885.

The birds of the old world, as well as the "two-legged creatures without feathers," are left behind by their more go-ahead American cousins.

There is a Californian Woodpecker which is not content with making holes in trees, after the manner of its kind on this side of the Atlantic, but corks them when made. There is a specimen of its work in the British Museum—a piece of the bark of a tree with round holes in it neatly stopped with acorns. It is not easy to say what the precise object of the bird in corking his holes may be, unless it is that he has stalled calves—grubs not quite at their best when first found—fattening in pens for future use.

Woodpeckers generally seem to be birds of an enquiring turn of mind. Among the curiosities of the Leyden Museum is the top of a telegraph post of hard teak, brought from Sumatra, with four or five deep holes drilled round the support of the wire by a little black and white fellow with a red cap, almost identical with our own London "Spotted Woodpecker."

His object in drilling the holes was, no doubt, to solve the mystery of the music of the wires, which seems as great a puzzle to four-legged creatures as it is to birds and children; for in parts of Norway much mischief is done to the telegraphs by bears, which, on the principle that "where there is smoke there is fire," take for granted that where there is a "hum" there must be a bee, and roll away the rocks piled up to keep the posts in their places to get at the hidden honey.

Perhaps the most remarkable feature of recent London ornithology has been the increase in the number of Wood Pigeons. When this sketch was first published, not very long ago, it was mentioned as worth noticing that the "deep mellow crush" of their note had begun to—

"Make music which sweetened the calm"

of Kensington Gardens, where one or two pairs had then lately settled. Now Wood Pigeons nest by dozens in all the parks, and it is a common thing to see fifteen or twenty together in one tree.

In our complex civilisation dangers to life and health crop up in such unexpected quarters that it is difficult to say where safety lies. Perhaps, though, the last of our London neighbours whom we should be inclined to suspect of dangerous proclivities, would be the masterless Pigeons, which swarm in all directions.

But a man may smile and be a villain, and birds are, apparently, no more to be trusted than men.

A lady lately took for a few months a house in Chester Square. The drains were duly inspected and pronounced faultless, and she took possession with every prospect of a pleasant season. It was not to be. A cloud of mystery hung over the house. Servants were disturbed by midnight rappings and awaked at daybreak by uncanny whisperings; and one after another complained of feeling ill, and gave warning.

When at last the lady herself had given way to the universal languor, and had, by doctor's advice, left town to seek fresh roses in country air, it was found that there was an unnoticed hole in the outer wall of the house, through which Pigeons had found their way in and out, and that the spaces between floorings and rafters were a big dovecote, evidently of several years' standing. There were living young birds snug in nests on guano beds under the floors, and dead birds in various stages of decay. Fourteen nests were found in the wall of one bedroom.

The origin of London tame Pigeons is lost in the mists of antiquity. Dean Gregory, in a paper on the subject, published in one of the church parish magazines, traces the colonies on St. Paul's Cathedral—of which there are two, one at the east, the other at the west end, which keep carefully apart, and it is said seldom or never intermarry—to the Fourteenth Century, when they were already well established. Among other authorities for this he quotes Robert de Braybrooke, Bishop of London, who, in 1385, wrote " there are those who, instigated by a malignant spirit, are busy to injure more than to profit, and throw from a distance and hurl stones arrows and various kinds of darts at the crows, pigeons, and other kinds of birds building their nests and sitting on the walls and openings of the church, and in doing so break the glass windows and stone images of the said church."

There is a legend that a hole was once neatly drilled in a window, and a bullet embedded in a book-case, within a few feet of the head of a high dignitary of Her Majesty's Civil Service, by a sporting young gentleman, who took a flying shot with a saloon pistol at a Pigeon in the quadrangle of Somerset House.

The Wood Pigeons are probably the only wild species of the *gallinaciæ*—the "poultry" order to which most of our gamebirds belong—common in London; but not long ago there was one very fine fellow to be seen in St. James's Park who deserves special mention. He was a cross between a cock Pheasant and a common hen, and had very nearly the head and neck of his father, with a half-dock tail; and could fly, if occasion required it, like a genuine rocketer.

In the next order, the "waders," we have Moorhens in plenty. In St. James's Park they are tame, and will scramble with the Ducks for bread from the bridge; but their habits are more natural in the Long Water. There one may watch them paddling about, jerking their tails or prying about shyly for what they can find on the grass outside the little cover by the water's edge. It is impossible to help believing that a Moorhen has an eye for natural beauties, and chooses the overhanging bough or fallen tree by the water for her nest, for picturesqueness quite as much as for convenience.

More than one Moorhen has been picked up on the premises of the Public Record Office, in Chancery Lane. It is not necessary to look far for the explanation, as the sky overhead is spider-webbed with telegraph wires running in every direction.

It is interesting to notice how soon resident birds learn the danger of the wires. When a line was first put up for a few miles along the coast from Cromer, Partridges, Woodcocks, and small birds—Larks particularly—were constantly picked up more or less mutilated; but, before the wires had been up many months, it was a rare thing to find a wounded bird.

Herons occasionally fly over London; but it is not likely that they often alight. Like most aboriginal tribes, they are gradually dwindling away before the progress of civilisation; and soon, if we wish to see them wild, we may have to go to the Dutch ditches or the unreclaimable swamps of America.

According to Michelet, whose delightful little book, "L'Oiseau," all bird lovers should read, the Heron knows he is the degenerate representative of a dethroned race of kings; and mopes in solitude, dreaming of the days of his glory, when his ancestors, the giant waders who left their footmarks in the secondary rocks, fought with great lizards and flying dragons, ages before a single mammal had appeared upon the earth. All the birds of which there are any very early traces were of the Heron tribe, and some of them must have been of enormous size. There are three-toed footprints in the red sandstone of the Connecticut* which are said to "measure

* The celebrated Connecticut "*Moulds*" are now believed "to have been made by certain extinct, in many respects, bird-like reptiles."—"The Elements of Ornithology."—*Mivart*

eighteen inches in length and nearly thirteen in breadth; and to indicate, by their distance apart in a straight line, a stride of six feet."

"They tell," says Hugh Miller, "of a time far removed into the by-past eternity, when great birds frequented by myriads the shores of a nameless lake, to wade in the shallows in quest of its mail-covered fishes of the ancient type, or long extinct molluscs; while reptiles, equally gigantic, and of still stranger proportions, haunted the neighbouring swamps; and when the same sun that shone on the tall moving forms beside the waters, and threw their long shadows across the red sands, lighted up the glades of deep forests, all of whose fantastic productions—tree, bush, and herb—have, even in their very species, long since passed away." There is no place in which the birds might be supposed to feel the change of times more than here. The Thames-side in old days must have been a paradise for long-legged birds; and even chaos itself and the modern world could be scarcely more unlike than the country round the little village of the Trinobantes, and the miles of brick and smoke two old Herns looked down upon, who flapped over London from the Essex marshes one day in August last.

It is told in a curious old book, called "Christ's Tears over Jerusalem," published in 1613, that at the time of the Plague of London, "the vulgar meniality concluded that the sickness was like to encrease because a Hernshaw sate (for a whole afternoon together) on the top of St. Peter's Church in Cornehill." But, adds the writer, "this is naught els but cleanly coined lies." There is a beautiful Heronry not many miles from London, well worth a visit, in Wanstead Park, the property of the City Corporation.

The ventriloquism of many birds, especially of the Heron and wild fowl tribes, is very strange. In the swampy districts of Finland, one may hear a party or Cranes apparently within easy shot, and with difficulty make them out almost invisible specks in the sky. Another morning, or very likely the same day —the projection of the voice seems to be independent of the state of the atmosphere—one hears what sounds a very distant cry, and is startled on looking up to see half-a-dozen great birds streaming along, not a hundred yards overhead.

The power, which is no doubt responsible for the legends common all over Europe, of spectral packs of hounds hunting the souls of the lost, is by no means confined to the high-flying birds. It is as impossible to tell from its cry where a Corncrake in a hay-field really may be as it is to guess the exact whereabouts of a passing flock of Geese.

There are probably men living still who have shot Snipe where Belgrave Square now stands. It is said that a very little time ago it was not uncommon to flush one in Hyde Park, between Victoria Gate and the Marble Arch; but the improvements of the last few years have probably banished them, at least till the days of Lord Macaulay's New Zealander. One reads occasionally of Woodcock picked up in the streets. A case of the kind was not long ago recorded in the *Field*. The poor bird had shared the fate of many of his kind, and had broken his wing against a telegraph wire in flying over at night.

Letters in the newspapers recorded that a Woodcock had flown by Buckingham Palace in the direction of Hyde Park at midday on the 21st October,

1884, and that in May of the same year a Dunlin had been seen feeding by the Serpentine.

Passing on to the sixth and last order, the webfooted, the ornamental waters in the parks are so well stocked with the different breeds of Ducks, that it is impossible to say to what extent they are frequented by genuine wild fowl. There is no doubt, though, that the number of occasional visitants is considerable; and, of those who are permanently quartered on the Serpentine, many fly strongly, and are, to all intents and purposes, wild birds.

Unlike most of us, their hours in London and the country are much the same. Flighting time—just as the last remains of the blurred red and blue which gives its peculiar picturesqueness to sunset in a big town is fading in the fog—is their favourite exercise time; and one may stand on the Serpentine bridge almost any autumn evening, and listen to Mallard and Widgeon whistling overhead, till, with a very small stretch of imagination, the Long Water becomes a tidal harbour, and the distant roar of Oxford Street changes into the break of the sea outside the sandhills.

In St. James's Park alone, besides black and white Swans, and ten sorts of Geese—four of them English: Brent, Bean, White-fronted, and Bernicle—there are, or were, a very few years ago, not less than nineteen or twenty distinct species of Ducks, with five or six crosses, including one beautiful one between the exquisite little Carolina and red-headed Pochard. About two-thirds are British, ranging in rarity from the Widgeon and Pochards—which still swarm in winter in the ponds and runlets in many parts of the coast—to the castaneous Ducks and delicately-pencilled Gadwall, one of the shiest and rarest of our English waterfowl.

The list includes, besides those already mentioned, Mallard, common Teal, and Garganeys, Shovellers, Pintails, the common Shelducks which breed in the rabbit-holes among the sand-hills by the sea, and the rarer "Ruddy" species, the tufted; and, perhaps most generally attractive of all, two or three Golden-eyes, with their brilliant blue-black and white plumage, and the eye, like a little drop of liquid gold, which gives them their name. They and the tufted and Red-headed Pochards are the life of the party, and are scarcely still for a moment together. It is amusing, in a general scramble for bread from the bridge, to watch them diving under the ruck, and popping up to snatch a crust from the very mouth of some sleepy fellow twice their own size, hunted in turn by half a dozen others as wide-awake as themselves.

Birds are many of them gifted with the lively imagination which can keep a child happily amused for an hour at a time with a cork on a bit of string for a dog, or "pretenting to be mother."

A year or two ago one of the Bernicle Geese in St. James's Park—no doubt with a history behind her and not improbably with a shot in the ovary to remind her of some "hair-breadth 'scape" on a frozen marsh in by-gone days—made a nest, and, without laying an egg, sat the regulation number of weeks on nothing more suggestive of goslings than the down from her own breast with which she had carefully lined it.

The next year, a second nest was made in the same spot, but this time the Providence which makes the woman to whom such family delights had seemed impossible to keep house and be a joyful mother of children, stepped in in the person of the keeper with a clutch of Ducks' eggs, which were safely hatched. For some reason or other the ducklings did not thrive — perhaps, because the old maid was too fussily anxious and gave them no peace, or, perhaps (as the keeper who watched them believed to be the case), because her sharp note jarred on ears which nature had designed for a mother's call in another key, frightening the poor little "boarded out" babies and making them restless—and before long the foster parents again were childless.

A couple of hundred years ago no one, with any pretence to education, would have been foolish enough to expect anything but failure in such an experiment as the park-keepers, Barnicles —" fowles like to wylde ghees which growen wonderly vpon trees " — being, as every one knew, the exception that proved the rule that birds are hatched from eggs.

The belief that the Bernicle Goose grew from the " pedunculated cirripede " that bears its name (*Lepas Anatifera*) lingered perhaps the longer because it was good for the Monks of Holy Isle and other northern monastries. "Men of relegyon" we are told in one of Caxton's priceless volumes, "eet Barnacles vpon fastynge days bycause they ben not engendered with flesche."

Hudibras made a slip in his natural history when he said that

> "Bernicles turn *Soland geese*
> In th' Islands of the Orcades"!

The first black Swans which were imported from

Australia could not at all understand the complication of the seasons which a change of hemispheres involved; and at Carshalton, one brood of little ones was hatched when snow was on the ground, unhappily only to survive in a handsome glass case. They have accommodated themselves to circumstances better now, and some fine broods have been brought up safely in St. James's Park. The Cygnets in the down are very like young white Swans.

A single Tern, noticed one blustering day a few winters ago, and a Stormpetrel, "Mother Carey's Chicken," reported to have been picked up alive in Kensington Gardens in December, 1886, introduce the "Longwings," the poetical family of the Albatross and Frigate bird. Unluckily, the Tern was some distance off, and the species could not be identified with perfect certainty; but a party of Kittiwakes who paid a well-timed visit to the Serpentine in 1869, when Mr. Sykes's "Sea-birds Preservation Bill" was under discussion, and other parties of the same beautiful birds, which have more than once since stopped for a time in one or other of the parks, have found themselves great people, and had all their movements chronicled in the fashionable news. The party which visited us in 1869 stayed some time, and were watched with pleasure by hundreds.

What Campbell wrote of the wild flowers is doubly true of the birds associated with the scenes of childhood. They can "wake forgotten affections," and "waft us to summers" and winters "of old;" and probably more than one old Londoner may have felt something not unlike a touch of home-sickness, and found his gas-dried eyes a little more moist than usual as he looked at their white breasts glancing in the sunshine. The tame Herring Gulls breed freely in St. James's Park.

On the 30th May, 1888, a Cormorant in full breeding plumage—white patches on cheek and thigh—appeared unexpectedly on the water in St. James's Park. He was first noticed by the keeper at half-past eight in the morning, and was tame and hungry enough to accept from him a couple of herrings for breakfast. A bird of the same species, no doubt the same, was seen a few days later on the Serpentine, and again flying over Lord's Cricket Ground in a northerly direction. "The bird," wrote Sir Ralph Payne-Galway, who recorded its last appearance in the *Times*, "flew fairly low, but owing, I presume, to Mr. Bonnor having just put a ball into the Pavilion, it escaped notice as far as I could judge, though it is true I heard one gentleman remark 'there goes a wild Duck.'"

Three Cormorants since imported from the Farne Islands have done well in St. James's Park, but have never yet bred or shewn any signs of an intention to breed. Some Guillemots and Puffins were brought at the same time, but, owing to the difficulty in procuring natural food, did not live long.

Of the last family of all, the Shortwings—the connecting link between birds and fishes—we have at times plenty of a single species, the little Grebe, "Dabchicks," lively little fellows, the quickest and best, perhaps, of our English divers, as much at home at the bottom as above the water. Of late years they have not been coming in such flocks as formerly, but in 1870 there were often as many as one hundred of them at once on the Round

Pond. They came and went unaccountably, and within a few days the place was alive with them and deserted again. As a rule, though, there were ten or a dozen at least to be seen feeding tolerably near the edge. They were then common, too, on the other waters in the parks. For the last few years six or seven pairs have bred regularly in St. James's Park. They commonly arrive late in March or early in April, and disappear with their families before the end of October. A nest built in 1887, in an exposed place, was, after it was finished, cut from its original moorings by the builders and towed a yard or two to a more secluded corner under an overhanging bush. Unluckily the second lashings were not so strong as they should have been, and, a fresh breeze springing up, the raft was wrecked and the four eggs it carried went to the bottom. After a sudden sharp frost in March, 1892, a Dabchick—a genuinely wild bird in good plumage—was found in a shallow puddle in the bed of the ornamental water, which had been run dry for cleaning, with one foot caught in the ice.

In the spring of 1883, after a spell of windy weather, a Willock—another of the "Brevipennes"—was caught alive in Russell Square. Why he came there, unless to prove his title to his other name, "the foolish Guillemot," it is not easy to say. It is a common thing to pick such birds up by twos and threes dead on the beach almost anywhere along the coast if it has been blowing hard on shore for any length of time.

They, and Razorbills, which, excepting in the form of the beak, are very much like them, are the commonest of the black and white birds which, on almost any voyage—northward more particularly—

one is sure to see sitting about in parties on the water, scarcely taking the trouble to lift themselves or do more than dive for a few moments as the steamer splashes by.

With the misguided Willock of Russell Square this chapter comes to a fit end, leaving us—such is the magic of every branch of natural history, under a sky clear from the pollutions of London dirt and London smoke.

The Birds of the Outer Farnes

"Thy tower, proud Bamborough, marked they there
King Ida's castle huge and square,
From its tall rock look grimly down,
And on the swelling ocean frown."—*Scott.*

MILLIONS of years ago, when the earth was still cooling and shrinking, and its crust every now and then wrinkling, like the scum on a saucepan of boiled milk not long taken off the fire, a great bubble rose from the depths and burst where Northumberland and Durham now lie. The explosion was felt from shore to shore on the mainland as it now exists and far out into the North Sea, and has left, among other memorials of its violence, the headland of once molten rock which has carried for centuries the magnificent pile of Bamborough Castle and the group of volcanic islands on which it looks down.

The Castle, after standing sieges innumerable and playing an important part in the turbulent politics of the Border, like Charles V. retiring to a monastery, has passed to a charitable trust. The fire-scarred basalt rocks from which its walls rise are in spring and summer pink and white with tufts of thrift and campion, and spotted at all sorts of corners with patches of another white, poetical only in the tale it tells of the domestic happiness of jackdaws and starlings beyond the reach of boys' fingers. The square central keep, when not occupied as a summer residence by some happy trustee, is let by the week for the benefit of the charity, and in the eastern wing thirty or forty orphan girls are housed and taught.

The Farne Islands, on which their bedroom windows look out, have a long history, too, of their own, scarcely second in interest to that of Lindisfarne or Iona itself. It was to the Farne, the principal island which gives to the group its name (one derivation makes it the "Place of Rest") that St. Cuthbert retired. It was here that he taught the eider duck the lesson of tameness during the breeding season, which she still remembers, though the drake, in common with most birds, has long since forgotten it; and here that Egfrid, King of Northumbria, and his nobles found the Saint, and on their bended knees, "with tears and entreaties," offered him the Bishopric of Hexham. It was on a rock on the Farnes that the *Forfarshire* went to pieces, and it is in the churchyard under the Castle on the mainland opposite that Grace Darling and her father sleep.

But for those whose calling obliges them to live more in the work-a-day present than in the past, the chief charm of the Farne Islands is that they are one of the principal breeding-places of sea birds on the English coast, and easily accessible from London. With the help of the Great Northern night express, a sleeping carriage, and fine weather, it is not difficult, at a pinch, to see all that is best worth seeing, and store one's memory with pictures not likely soon to fade, without being away from Pall Mall more than a day.

The best time to visit the islands is usually about the last week of May or first week of June, to see eggs; or, to see the young birds, three weeks or a month later. It was not until the 14th of June that we were able to make the trip, but owing to the lateness of the season we found ourselves early enough to see the eggs in perfection, scarcely any of the birds having hatched off.

When we had arrived at Bamborough the afternoon before the weather had not been encouraging. It was blowing a quarter of a gale, with heavy thunder showers; but in the evening the sky had cleared a little and the sun found its way through the clouds, to set in a wild confusion of banked reds, yellows, and purples. We woke to find the morning bright, and by the time we had breakfasted and found our way to North Sunderland, three miles off, where a boat was awaiting us, the wind had died away, and the only fault, if any fault could be found with the day, was that there was scarcely breeze enough for sailing.

Our object being to see as much as we could of the birds, and opportunities uncertain, as threatening clouds manœuvred still on the horizon, we steered at once for the Outer Islands, the chief nesting places, leaving a mile or two to the left the inner group, which are well worth a special visit:—Farne, with its chapels and its "churn," a rock-bridged cleft, through which at half-tide, when the wind is blowing heavily from the north, the sea is said to spout in columns ninety feet high, a statement the truth of which we were happily unable to test for ourselves; the two "Wide-opens"; the "Scar Cars"; and four or five others with names as uncouth, corruptions most of them of Anglo-Saxon* descriptive titles.

Terns and gulls had been from the time we started hovering round us singly or in twos and threes, and an occasional guillemot or puffin had dived out of the way of the boat or risen with trailing splash and the sharp quick beat which is characteristic of the flight of short-winged birds; but it was not until we had been afloat for an hour or so, and were nearing the Brownsman, our first landing-place, with the Crumstone and Fang on our right, that we had any taste of what was to come.

The whitewashed tops of the black basaltic rocks which faced us shone in the sunshine, and through a glass we could see they were lined, without a gap, with motionless figures, looking in the distance like an army of dwarfs, in black, with white facings, drawn up in review order to receive us. As we pulled into a little bay, hidden from us until we rounded a corner by the Gun Rock, we found ourselves the centre of a startled screaming multitude of puffins, gulls, and terns, and a few minutes later ran the boat aground, and landed on the slippery rocks.

In early times the knowledge that the birds which took sanctuary on the Islands were under the miraculous protection of St. Cuthbert was security enough for them and their eggs. " Beatus etenim Cuthbertus," wrote Reginald of Coldingham in the reign of King Stephen, " talem eis pacis quietudinem

---

* A table, giving in parallel columns the names in the forms in which they appear in records stretching back seven or eight hundred years at least, will be found, with much interesting information on other matters, in a monograph on the Farne Islands, by Mr. George Tate, published in 1857 by the Berwickshire Naturalists' Society.

præbuit, quod nullus hactenus hominum eam impune temerare præsumpsit."

Once on a time an unlucky monk—Leving, servant of Elric the hermit, uncle of Bernard, sacrist of Durham—in a moment of weakness, when his holy master was away, yielding to his lower appetite, killed a duck and ate it, scattering the bones and feathers over the cliff. When, fifteen days later, Elric came back he found bones, feathers, beak, and toes, neatly rolled up into a parcel—" cunctis in unum convolutis "—and laid inside the chapel door. " The very sea," says the devout historian,* who had the tale first hand from the repentant monk, "not having presumed to make itself participator in the crime by swallowing them up." Leving was flogged, and for many years—though there are records of puffins and other "wyelfoyle" sent from the brethren on the Farnes as delicacies for high-day feasts at Durham— St. Cuthbert's peace was probably unbroken.

But saints in these freethinking days have lost something of their power, and need at times, to enforce obedience to their commands, the help of the secular arm, and a year or so ago it somehow or other came to pass that the birds found themselves practically unprotected in any way. The nests were at the mercy of anyone who cared to land, and were robbed so recklessly that the extinction of the colonies was threatened. The danger has happily this year been met by the public spirit of a party of philornithic gentlemen, who, with Mr. Hugh Barclay, of Colney Hall, Norfolk, at their head, have leapt into the breach and obtained a lease of both groups

* " Reginaldi Monachi Dunelmensis libellus de Admirandis Beati Cuthberti virtutibus," cap. xxvii. (Published by the Surtees Society in 1835.)

of the Farnes. They have placed at their own cost watchers on the chief islands, and give leave to land to anyone who promises in writing to conform to the rules of their association, one of which is that without special permission not a single egg shall for a time be taken.

What most forcibly impresses a visitor on landing, after he has recovered a little from his astonishment at the number of birds still remaining and their tameness, and his ears are becoming more accustomed to the Babel of cries all round him, is perhaps the regular and orderly manner in which the nesting-grounds are divided among the different species, and the honourable manner in which the arrangements agreed upon are carried out. According to Reginald it was St. Cuthbert himself who mapped the Islands out for them.

The first colony we invaded consisted entirely of the lesser black-backed and herring gulls. These two species (the black-backs were by far the more numerous, perhaps in the proportion of eight or ten to one) share between them the flat table-land of the island, which is patched with a thick growth of bladder campion and another plant, with a succulent stalk and white blossom, but for the most part bare rock, split into steps, with little but lichen growing on it. The nests, which are placed without any attempt at concealment, are all on the ground, and are at best a few stalks of grass or campion arranged like a saucer, but in many instances the eggs are laid without even this provision being made for them. They were as thick on the bare rock as in the cover. One or two nests had in them young birds in speckled down, just hatched; but nearly all had two or three eggs in, varying often much in colour.

The eggs of the two allied species breeding together can be distinguished only by marking the nests as the birds rise. It is a peculiarity of the gulls generally that eggs are often laid after the bird has begun to sit, and it is a common thing to find eggs fresh and hard set in the same nest.

But the most curious sight on the Brownsman Island was the adjoining colony of the guillemots. These, so far as we saw then, were entirely confined to the tops of the Pinacle Rocks, which had first attracted our notice. Stray birds, we were told, occasionally breed in other parts of the island ; but we saw no eggs elsewhere. The Pinacles are three or four precipitous columns of black basalt, inaccessible except by ladders, separated from the mainland of the island and from each other by narrow chasms running sheer down to the sea. The tops are flat, and as we stood on the edge of the rocky cliff, opposite and on a level with them, we saw at a distance of only a few yards masses of guillemots, most of them, so far as we could see, sitting, or rather, it seemed, standing, on an egg, and wedged together as closely as sheep in a pen.

A few had the white lines round the eyes—like spectacles—which is the distinguishing mark of the rarer "ringed" or "bridled" variety ; but almost all were the common bird well known, in winter especially, on every part of the coast. It would be impossible to form any estimate of the number we looked down upon ; but, in spite of the attraction of a shoal of small fry of some kind a mile or so out, which was the centre of interest to an excited white and grey cloud of birds and must have thinned considerably the party at home, there could not have been less than several thousands on the rocks. A field-glass carried us into the middle of the crowd, and we could see all they were doing, and almost fancy we could hear what they were saying and read their characters. Some of the matrons—probably it was not their first experience of the breeding season—looked intensely bored. They reached out first one wing then another, gaped, got up for a moment and stretched themselves, and yawned again, with ludicrously human expression, conscious evidently of what society expected from them, and submitting to its restraints, but heartily sick of the whole concern, and longing for the time they might be free again to follow herrings and sprats at their own sweet will, without haunting visions of a chilling egg.

Others seemed entirely absorbed in their eggs. There was one bird in particular which we watched for some time, the proud possessor of a brilliant green, strongly marked egg—as usual to all appearance quite out of proportion to her own size—which she arranged and rearranged under her, trying with beak and wing to tuck the sharp end between her legs, but never quite satisfied that it was covered as it should be. But for the wonderful provision for its safety in the shape of the guillemot's egg (a round flat-sided wedge, which makes it when pushed turn round on the point, instead of rolling, as eggs of the usual form if placed on a bare rock would do), most of those we saw would probably have been dashed to pieces long before.

It was an old belief[*] that the eggs of such cliff-

---

[*] " Locus nempe, (ut dixi) cœmento albo incrustatur, ovumque cum nascitur lentâ et viscosâ madet humiditate quâ cito concrescente, tanquam ferrumine quodam substrato saxo agglutinatur."—*Harvey, De Generatione Animaliorum.*

haunting birds when first laid were coated with a natural glue which hardening at once fixed them to the rock.

As is commonly the case with basaltic rocks, the precipitous faces of the Pinacles and the cliffs opposite are lined with cracks running across and up and down, and broken into steps and shelves accessible only to birds or the boldest trained climbers. These, with the exception of a few of the larger upper ledges, which go with the tops of the Pinacles, and are part of the family estates of the guillemots, are tenanted by kittiwakes. Their nests, which are also of grasses or dry seaweed, and occupy all the most tempting corners, are much more carefully and substantially built than those of the larger and noisier cousins on the table-land of the island, and the bird, as she sits snugly—" coiled up," perhaps, best describes the favourite attitude—on her eggs, with her white breast exposed and head turned over her shoulder, the yellow beak half hidden in the pale blue feathers of her back, or raised only for a moment as her mate sails up with the last bit of gossip from the outside world, looks the perfection of peace and comfort, the greatest contrast imaginable to the uncomfortable Babel of the guillemots, a few feet above her. The eggs, like those of most sea-birds, vary much, but are, perhaps, proportionately shorter and thicker than those of most gulls, and have usually a ground colour of greyish green. Four or five eggs is not an uncommon number for a kittiwake to sit upon; but none of the nests into which we were able to look had more than three in it.

As we passed a clump of campion on our way back to the boat, we all but trod on an eider duck, who was sitting on a couple of eggs. She rose slowly and heavily, with a flight like a greyhen's, and lit a few hundred yards out to sea, where she was at once joined by her handsome mate, who had been concealed on guard not far off among the rocks of the bay. The drake—unlike the duck, which, when nesting, entirely changes her habits, and becomes, as we saw for ourselves, as tame as an Aylesbury, allowing herself to be almost touched before she rises—never loses his habitual wariness. He is seldom far from the duck, but, excepting as she leaves her nest, when he is pretty sure to join her, manages to keep well out of sight. They are very common on the islands. We saw a great many nests, several thickly padded with down, but—perhaps because the black-backed gulls are bad neighbours, as sucked egg-shells here and there too plainly showed—none had larger clutches than four or five. One forgiving duck was sitting on two eggs, one of which was a gull's.

The eider duck, when frightened, usually, as she rises, spatters her eggs with a yellow oil, which has a strong, sickly, musky smell. The young birds are taken by their mothers to the sea almost immediately that they are hatched; but we were lucky enough, later in the day, on another island, to find, under a piece of stranded wreck, four tiny brown-black ducklings. They were not many minutes out of the shell, and looked, in their soft bed of down, which exactly matched their own colour, the perfection of baby comfort. One of the watchers had noticed eggs in the nest an hour before we found the little birds.

From the Brownsman we crossed to the South Wawmses, which, with its sister island, the North Wawmses, from which it is separated by a narrow channel, is the headquarters of the puffins. We landed in a shingly creek, and as we climbed the rocks, which are here rather a bank than a cliff, we were met by a string of startled puffins, which came with quick, arrowy flight, straight at us, passing out to sea within a foot or two of us. The rocky foundation of both Wawmses is covered in parts with a dry, light peat, which is honeycombed in every direction with burrows, most of them containing one very dirty white egg, protected in many cases by the parent bird, which, when we put our hands in, fought with foot and bill, biting sometimes hard enough to break the skin and draw blood. We drew one or two birds out of their holes. They fought to the last, and when we let them go, more than one waddled back to her treasure, with an indignant shake and look which said very plainly, "I've taught that fellow a lesson he won't forget in a hurry."

There is something irresistibly comic in a puffin on his native soil. With his little round body poised straight on end on turned-out toes, and impossibly coloured beak, which does not seem really to belong to the face at all, and his grave earnest expression, the bird looks like nothing so much as a child with a false nose on, dressed in his father's coat, playing at being grown up. They are on another ground very interesting birds. With comparatively few exceptions, when birds build in holes, where colouring is unnecessary for purposes of concealment—kingfishers, woodpeckers, and petrels, for instance—they lay white eggs. When they lay on the ground in the open the eggs are coloured, often in such close imitation of their surroundings that one may pass within a foot or two without noticing them. We saw on the Farne Islands tern's eggs among the stones, and ringed plovers' eggs on the sand, so exactly matching the ground that, though we looked closely, with the certainty that eggs were near us, it took some time to find them.

We cannot tell how many thousand generations back it was that the ancestors of the puffins of our day came to the conclusion that burrows were the best places for the family to breed in, but, in the matter of egg-painting, they are still apparently in a transitional stage. The eggs, when not too dirty to show their natural colour, are almost white, but at the thick end there are usually faint spots, just sufficient to show that, though the painter's art has been long neglected, the brushes are there, and the internal colour-box has still a little paint in it, and might, if a change of tastes at some far future time required it, be filled again.

While we were amusing ourselves with the puffins on the Wawmses, a fresh breeze had sprung up, and as soon as we had finished luncheon we hoisted a sail, and after landing again for a minute on the Brownsman, which we had first visited, to look for a nest of the rock pippet, which is rare in more southerly parts, but breeds here plentifully in the grass tufts in the cracks of the rocks, sailed across the Sound to the Wide-opens, which we had passed without landing in the morning. The Wide-opens—once "Weddums," the "Ragers"—had in early days a very bad reputation. It was to them that St. Cuthbert banished the devils which, when he first came to Farne, had annoyed him very much, and after his death became again so bold that they took no trouble to conceal themselves, and were a constant anxiety to the monks on the neighbouring island.

We were received ourselves with screams as we landed, but of a note less alarming than those which night after night, kept the good saint's successors awake. The sunshine was broken by clouds of terns, perhaps the most exquisitely graceful forms of bird-life, and, as we looked to our feet to avoid treading on their eggs, which lay thickly strewn on the ground, little black shadows with forked tails and wings crossed and recrossed, circling backwards and forwards on the sand.

Four kinds of terns—the "common" and the "Arctic," from which it is scarcely distinguishable; the "Roseate" and larger, black-billed, "Sandwich" tern—breed in numbers on the Wide-opens. We had met with a few stray eggs of the "common" or "Arctic" species—without catching the bird on the nest, it is quite impossible to say to which of the two an egg belongs—on the other islands; but they were nothing compared with the numbers we now saw. It was the eggs of the Sandwich tern which we wished more particularly to see. They are very large for the size of the bird, and unusually boldly marked. Though there is no difficulty in recognising them at a glance, they vary infinitely, no two being painted exactly alike. We found them collected together (probably to the number of several hundreds) among the sand and shingle-heaps on the higher grounds, usually two or three in a nest. The Sandwich tern is said to be much more easily frightened than either the "common" or "Arctic," and, if harassed during the breeding-season, changes its nesting-place, often quite deserting an island. A few years ago the bird was much more plentiful than it now is on the Farne group; but happily the colony on the Wide-opens shows as yet no sign of early extinction.

Within a few hundred yards of us was the House Island, with its historic buildings; but a fine day, with surroundings such as ours had been since we started in the morning, slips by very quickly. The Megstone Rocks lay a mile or two off, and we could not miss them. If we were to catch the night express at Belford, either dinner or the ruins must be sacrificed, and to have hesitated in our choice would have been an insult to the keen air of Northumberland.

The "Megstones" are bare volcanic rocks, with no vegetation on them but the seaweeds below high-water mark and an occasional patch of lichen. The chief rock is a breeding-place of cormorants, no other birds apparently venturing near it. A ship had a few weeks before our visit been wrecked on the rock. The solitude had been for some time disturbed, and we were warned not to expect to see much, but as we neared the rock we saw heads on snake-like necks stretched up here and there, and as we watched our opportunity to spring from the boat a black cloud of cormorants rose together within a few feet of us.

Of the many allusions to birds to be found in Milton's poems, there is scarcely one which is not more suggestive of the study than of the open air. But there is an exception. The idea that Satan when he first broke into Paradise, and wished to look round him unobserved, got on to the Tree of Life, and there "sat like a cormorant devising death," must have been taken first-hand from Nature, stored up, perhaps, for future use in the days when the poet, on leaving Cambridge, with eyes not yet "with dim suffusion veiled," made his voyage to the Continent. There is something diabolical in the pitiless cold glitter of the green eye over the long hooked beak, from which the most slippery fish, once seized, has no chance of escape, and the distinctly sulphureous smell of its haunts is in keeping with the look of the bird.

The cormorant has for some wise reason (perhaps to help its rapid digestion, or perhaps to neutralise to some extent the smell of stinking fish—if the latter is the intention the work is very poorly done) been gifted with an extraordinary power of secreting lime. The entire surface of the Megstones for some distance round the nests—of which we counted ninety-three, almost all with eggs in—looked as if it had been freshly whitewashed. The eggs are long and narrow, without much difference between the two ends, and if held up to the light and looked at from the inside through a hole are beautiful, many of them being as green as an emerald or as the eye of the bird itself. But seen from the outside they look like eggs which a boy has begun to cut out of a lump of chalk, and left only half finished, irregular blotches of rough lime sticking out on many of them.

The nests are round, and built of dry seaweed. They are about two feet across, or a few inches more, and many of them not much less in height, and built with great regularity, looking almost as if they were lengths cut from a black marble column, slightly cupped at the tops, and, curiously enough, stood out most of them from the whitewashed platforms unspotted.

The only other sign of life which we saw on the Megstones did not detract from its lonely wildness. It was a long-legged, thin, wild-looking blackbeetle, which had been sunning itself on the hot rock nearest the highest point. It rushed towards us, as if to attack, at a great pace, and before we could catch or identify it threw itself over a precipice and escaped into a crack at the bottom.

The wind was fair for the shore, and as the water lapped our bows the Megstone Rocks settled down fast, lower and lower, into the sea behind us. The turrets and battlements of Bamborough Castle, which seen on end recalls the Normandy St. Michael's Mount, separated themselves one by one from the block, and sooner than we could have wished, we were landed safely a mile or so from the village on a natural jetty of rock, at the end of which we had watched the evening before an eider drake addressing, with much gesticulation, a party of ducks. A few hours later we were comfortably asleep, rushing through the night to London.

Of all the poor creatures whose fate it was to be strangled or battered to death by Hercules, there was only one who made a really good stand-up fight, and at one time seemed to be fairly beating him. He was Antæus, the son of the Earth.

Every time that he fell and touched his mother—we should say, " ran down to the country "—he came up again with fighting powers renewed. It was not till Hercules found out his secret and held him up, never letting him fall—we should say, " stopped his Saturdays till Mondays out of town "—that he quite broke him down. It is a myth in which the wisdom of the ancients has written for our admonition, on whom the ends of the world have come, the lesson that the best cure for a tired head and irritable nerves is the touch of Mother Nature—to escape from the rattle of cabs and omnibuses, and the everlasting cry of "extra specials," and lose oneself, if only for a day, among the wild creation.

Nowhere in the languid days of early summer—

the breeding season of the sea birds—can the tonic be drunk in a pleasanter or more invigorating draught, than on the rocks and islands of the Outer Farnes.

The Shetlands

> ". . . . The living clouds on clouds arise,
> Infinite wings! till all the plume dark air
> And rude resounding shore are one wild cry."
>
> *Thomson.*

THERE is a story of a little boy who used to feel sick when he sat in a carriage with his back to the horses. So long as he was small enough to sit on his mother's knee, or as a third on the front seat without crushing his sister's frock and making her a figure, his weakness did not much signify. But when he grew too big for this, his mother told him he must try to be a man, and get over it. He wished to please her; and, having a fairy godmother who helped him when she saw he was trying in earnest, succeeded so well, that soon he had learned to travel backwards as no other boy before or since has done. Often he would shut his eyes and spin back at first for hundreds, and then, as he grew more accustomed to it, thousands of years, until one very hot steaming day as it seemed to him—though at home it was cold enough for a fire in the schoolroom—as he skirted, with boots very wet with red mud, a wood of overgrown mares' tales, he nearly trod on a Pterodactyl, which he had not noticed in a reed bed till he was close by it. It snapped at him as it rose at his feet and frightened him. After that, excepting in his mother's carriage, and sometimes in the train, he would not go backwards any more, but began to go forward instead, and when he went to school was soon head of his form.

The feelings of the little boy in the story when in his backward journeys he found himself with palæozoic surroundings, must have differed in degree only from our own when, on Whit-Tuesday last, with the din of London scarcely out of our ears, and recollections of flowers and uniforms and ladies' dresses on the Foreign Office stairs fresh in our minds, we found ourselves on a remote promontory in Shetland face to face with living examples of life, under circumstances which almost everywhere else in the British Islands have long since passed away.

The green of the turf at our feet was broken with patches of thrift and pink campion, and starred in all directions with dwarfed blue squills in full blossom. On the opposite side of the Sound, to our left as we looked southwards, a mile or so off, lay the Island of Mousa, with its almost perfect Broch in full view. To our right lay a little land-locked bay, a perfect anchorage for a Viking's boats, with deep water still as a pond, though a stiff breeze was blowing, and both open sea and Sound were white with breakers. On the narrowest point of the isthmus were the ruins of a second Broch commanding the promontory and bay; and on the mainland opposite, within twenty yards, stood a crofter's homestead, built with stones from the Broch, not many degrees removed from the bee-hive huts, of which the outlines, and in more than one case the stone foundation walls, clustering round the castle, were still to be seen.

We leant against a corn tub with a roughly chipped disk of stone for lid, which might have passed muster in a museum as a relic of prehistoric days, and chatted with a kindly old lady, wearing "revlins," the most primitive form of shoe known, made of untanned cowhide with the hair on, fitted to the foot while "green," to the use of which, writes Professor Mitchell, "John Elder referred in his famous letters to Henry VIII. of England (1542-43), when he wished to show the extent of barbarism of the 'Wilde Scotes.'"

We had surprised her by expressing a wish to see a quern in working order, and she took us through a gate, swinging on a stone socket, into an outhouse to see one belonging to her uncle and herself. The door was so low and the walls so thick that we had to stoop almost to "all fours" to get in, and having done so, found ourselves in the dark until our hostess had found her stick—a precious possession where there is no native grown wood—and opened the shutter by knocking off a sod which covered the only window, a slit in the turf roof. The sun at the moment being clouded, and the light, even when the shutter was down, not very brilliant, our friend left us to fetch a lamp. We were quite prepared to see her return with a Shetland "Collie"—the double iron pan with pointed spouts like a jug (the one to carry the melted blubber and wick, the other to catch the drip) which, until whale oil gave way to paraffin, was the common lamp of the country—and were almost disappointed when, instead, she brought a contrivance of scarcely less primitive design, not unlike a battered tin teapot with a twist of unspun wool in the spout for wick. In spite of the cloud of smoke it threw up, and the rather troublesome attentions of a small calf which had been shut up in the room to keep it from its mother, we were able, by the light it gave, to examine, underneath the wooden tray on legs, fastened to the wall, on which the grindstones were fixed, the simple but very effective contrivance* for regulating the coarseness of the meal to be ground.

We felt as we crept back into the open air much as we might have done if, on crawling down the rocks outside to look for the nests of the black Guillemots which swarmed on the lower ledges, we had turned a corner and come upon a great Auk sitting on her egg.

Perhaps the sense of far-backness was all the stronger upon us, because, since we had left London, a veil had been dropped between us and our past existence. The weather as we left Aberdeen had been perfection, with just enough air stirring to freshen the colours of the sea, and carry the smoke of the funnel clear of the deck. The sun set "smilingly forsworn," at twenty-five minutes to nine, and as the long twilight, which brought home to us that we were getting northward, set in, sheerwaters—which in their habits are the owls of the sea, living for the most part in their holes on shore by day, and coming out at dusk—shot past us, one or two at a time, with quick gliding flight, on their way to their feeding-grounds, the long, sharp wings closing at each stroke backwards, until the birds seemed to have forked tails like Swallows.

Perhaps if our experience of local weather signs

* A full description of the mechanism of a quern, with illustrations, with much other interesting information with regard to the survival in Shetland of implements, &c., of patterns of very early date, will be found in the Rhind Lectures, delivered in 1876 and 1878, by Dr. A. Mitchell, Professor of Ancient History to the Royal Scottish Academy, quoted above, published in 1880, under the title " The Past in the Present."

had been larger we might have seen a warning of what was before us in the curiously angular shape of the sun as it dipped; but ignorance was bliss, and we "turned in," happy in what we thought the certain prospect of a quick and pleasant voyage, and woke to find ourselves anchored for five-and-twenty hours in a dripping fog, somewhere near, but no one could say how far from, Kirkwall Bay.

The interest of our trip lay more in the present than the past; our object in coming so far having been not so much to look for antiquities as to see the birds, which in the summer gather by myriads to breed on the rocks and islands of the Shetlands. Some which are common here, nest in few, if any, other places in the British Isles. When we started we had indulged in dreams of visits to Fair Island, and perhaps to Foula, which lie, the one—reported to be more beautiful than any island in either Orkney or Shetland—half-way between the two groups; the other—the wildest and most precipitous in either—in the open Atlantic, some twenty miles or so to the west of the mainland of Shetland.

But twelve days, or at most a fortnight, was all that we could conveniently spare, and of these three had already gone before we set foot on shore in Lerwick on Sunday evening.

It is only in very calm days that a landing can be effected on either Fair Island or Foula, and as the weather, which for the fortnight before our arrival had been unusually warm and still for the time of year, had broken, and the "Beltane Ree," of which before leaving home we had read with some misgivings in Dr. Edmundston's "Glossary of Shetland Words," as "a track of stormy weather common in the Islands about Whitsuntide," was to all appearance upon us, we were obliged to give up all notions of anything more ambitious than a visit to one or two of the more easily accessible of the chief breeding places, and to the castle of Mousa, which we were especially anxious to see.

The welcome breeze which had blown away the fog, had, since it first sprang up on Whit-Sunday, been steadily freshening, and by Monday morning, when we started for Noss, an island lying outside Bressa, half a gale was blowing.

It was some little time before we succeeded in getting a boat to carry us over the Sound, but at last one was found, and by eleven o'clock we were landed on the other side, with luncheon in our pockets and clothes comparatively dry. A pleasant walk of three or four miles leads from the landing-place to the point of Bressa, opposite the shepherd's house in Noss, where there is a ferry between the two islands; and half-way across, as we sauntered along, interested by such un-Londonish sights as women harnessed to harrows, or carrying heavy loads of peat from the hills in straw baskets hanging from their shoulders, knitting as they went, we were delighted at seeing for the first time, near a freshwater lake, a party of Richardson's Skuas—the birds which more than any others were responsible for bringing us over land and sea eight hundred miles and more from London. We knew that they bred regularly in Mousa, some fifteen miles to the south, and on some of the more northerly islands, but had not expected to find them in Bressa or Noss; and the first sight of their long, thin, sharp-cut, angular wings, and the two unmistakable long pin feathers springing from the middle of the tail, and the powerful, graceful flight of the birds as they circled round, playfully chasing one another, or lit on the water to rise again the next moment, had the charm of a welcome surprise.

Noss is separated from the larger island by a narrow cut. The channel is not many yards wide, but in certain states of wind and tide cannot be crossed without danger. We had been warned in Lerwick, that with the wind blowing as it had done for some thirty hours, it was not unlikely that we might find the ferry too rough to cross. But this time fortune favoured us, and though the sea close by to the north was white and thundered ominously, we got over without any difficulty.

From the landing-place, where the shepherd's cottage, the only one on the island, stands on level ground not many feet above the sea, the land in Noss rises westward, steepening at first gently, then more and more rapidly, until, at the split-sugar-loaf-shaped point of "the Noup," the short flowery turf ends abruptly in a precipice.

Not far from the highest point is one of the many little inlets known locally as Geos, Voes, or Wicks, according to their size and shape, which give much of its picturesqueness to the coast scenery of the Shetlands; and from the promontory at the farther side we were able to get a good view of one side of the sea face, which seems to be built up of thin horizontal layers of sandstones and conglomerates, alternately hard and soft, which, weathering with curious regularity, have given the precipice in parts very much the appearance of a gigantic bookcase, on every shelf of which, as we saw it, were tightly packed masses of sea-birds, of every shade of white, black, and grey.

We had been told that to do justice to the Noup of Noss, it should be seen from the sea. It may be so. But if the view from below is more impressive than that on which we looked down from the summit, it must be one of extraordinary grandeur.

The waves were rolling in, and breaking into foam on the rocks 600 feet below us. Puffins, Guillemots and Shags shot in and out by thousands. Gulls in numbers incalculable sailed round and round or hung motionless in the wind—so near some of them that, without any need for glasses, we could see the ruffling of each little feather, and the expression of eyes turned on us—and faded in perspective as we looked down into a living milky way of birds.

To make the picture complete, a Peregrine Falcon, monarch—in the absence of the white-tailed Eagles, which have usually an eyrie either on Noss or Bressa —of all he surveyed, looking, far up in the blue scarcely bigger than a fly, screamed in notes, which rung out clearly above all other sounds, defiance to the world at large.

Nothing that ever has been or ever will be written of such scenes, will make the reader see them with his own eyes for the first time, or for that matter for the hundredth, without a sense of almost dazing amazement at the numbers in which the birds collect.

A couple of hundred yards or so from the southwest of the Noup, lies the "Holm," a corner of the main island, cut off by a chasm, through which the sea runs. The Holm some years ago was connected with Noss by a rope bridge, put up by a reckless cragsman who lost his life on returning after the work was completed. It is now inaccessible, and was, when we saw it, crowded with nests of the lesser black-backed and herring Gulls, which here, as elsewhere, breed socially together.

Birds'-Nesting Season.

In the remoter islands something of the old spirit of the Norseman, who believed that the only safe road to Valhalla was across a bloody battlefield, still survives in the idea that the most honourable deathbed for a Shetlander is "on the Banks;" but on the more comfortable mainland, so far as we could learn, there is very little cliff-climbing done now by any but adventurous boys ; and, excepting when, as hundreds are misguided enough to do, the birds tempt fate by laying on the flat, they may most of them reckon on bringing up their families without human interference.

As we stood by the Holm, continuous flocks of small Gulls, either Kittiwakes or Sea-mews—the two are in appearance so much alike, that unless very close indeed it is impossible to say which is which—flew over us, all in the same direction, coming from the north-west. Every bird, in all many hundreds, had a bunch of something in its mouth. We tried to find out what the attractive morsels were, but all our efforts to make one of them drop his load were useless, and we could only guess from the general appearance and size (very likely wrongly) that they were parcels of sand eels or sand worms.

From the Holm we strolled over to the lower ground, where in the morning we had noticed more than one anxious pair of Richardson's Skuas, and were absorbed for the rest of the afternoon in watching them. The Skuas, of which there are four kinds classed as British, are the connecting link between the Gulls and Hawks. The Richardson or Arctic Skua is the commonest. It is a slender bird with a body scarcely bigger than a Pigeon, but with a powerful cutting beak, and great powers of flight and courage. They live, like all their tribes, almost entirely by robbing larger Gulls, and fly at birds three times their own weight and size as fearlessly as a Sparrowkawk flies at a Lark.

As we lay on the side of the hill, looking down on the hollows which are their favourite breeding places (they make no nest), a Skua, for no other reason apparently than that our continued presence too near its eggs had put it out of temper, dashed savagely at a Gull which looked nearly big enough to swallow it, and struck it now from above and now from below with a crack which sounded as if the blow had been given with a riding-whip. The poor bird attacked made one or two attempts to get back to the two eggs in a nest on the grass beneath us, from which just before we had driven it, which was all it wished to do, but in the end had to give it up as a bad job, and flew off with a protesting wail.

There is nothing in Nature more beautiful than the "heaven taught art" with which most birds which breed on the ground in the open lead away from their eggs and young. The Oyster-catcher (perhaps because he feels that it is hopeless for a bird dressed in staring shepherd's plaid, with red legs and beak, to hope to conceal himself) loses his head completely, and betrays his nest by shrieking despairingly over it the moment it is approached. But he is only the exception which proves the rule. We saw in one place, within a yard or two of our feet, what looked like a sand-coloured mouse, crawling slowly and stealthily close to the ground, down a little hollow, following the indentations of the ground where the sand, which had drifted between tussocks of grass, exactly matched its colour. It was a little ringed Plover, afraid, if it rose as shyly as at any other time it would have done, of betraying four pointed eggs, evidently hard set, arranged, points inward as a

Birds'-Nesting Season.

Maltese cross, in a saucerful of little scraps of sandstone and speckled granite, carefully chosen to match their colouring.

But for the knowledge that almost all birds, if their nests are disturbed at all early in the season, lay again,* the prick of conscience, without which an egg which the bird has been at so much pains to conceal cannot be taken, would be too dear a price to pay, even for the pleasure and interest of a collection, with the refreshing recollections it can awake of "thick groves and tangled streams" hunted in boyish days, and island-dotted lakes, moors and marshes, and seabeaten headlands, since visited in intervals of sterner occupations.

Most Sea-gulls, certainly the herring and lesser black-backs, whose eggs are largely collected for food wherever they are at all common and easily got at, have very considerable powers of egg-production at will, though the ordinary "clutch" when undisturbed is seldom more than three or at most four. The only difficulty seems to be with the colouring material, which is apt to run short, and the more eggs are taken, the paler, as a rule, becomes the ground colour and the less clear the markings.

It is a fairly safe assumption that an egg unusually strongly marked or highly coloured is one of the first of the season which the bird has laid, and it is not an uncommon thing, at least with Gulls, to see the pitch of colour in a nest containing one or more of such smart eggs brought down to the average by an unusually pale egg or two in the same nest.

The Scoutie ailen, as the Richardson's Skua is called

* A remarkable instance of the perseverance with which a bird will at times cling to the spot selected for a nest is recorded in Appendix B.

in Shetland, carries the ordinary arts of deception to as great perfection as any bird. It can limp like a Partridge, and drop as if shot from the sky, and lie on its side feebly flapping one wing. But if the stories told by the shepherds are true, and certainly our own experiences strongly confirmed them, the bird is not content with such tame devices as these.

In Flaubert's wonderful book " Salammbô," when Hamilcar learns that, as a last hope for the city, a sacrifice of first-born to Moloch has been decreed, he hides the little Hannibal in dirty clothes in the slaves' quarters, and struggles with the priests, who tear from his arms a jewelled and scented slave boy.

The Scoutie, with the true spirit of the noble Carthaginian slaveowner, when hard pressed, deliberately leads on to the nest of the Gulls it despises, and then goes through an elaborate pantomime of distress. Again and again we made sure that at last we were to see the true Skua's eggs, and as often found ourselves looking at the nest of some common Gull.

But, before returning to Lerwick, we were to be treated to an even more amusing specimen of the cynical humour of the Scoutie. One of our party had for some time watched a bird, which evidently had eggs close by, and at last, when its suspicions seemed to be lulled to sleep, saw it light on a rough spot not very far off. There it stopped in ostentatious concealment, every now and then cautiously lifting its head and peering over the grass in his direction. He marked the spot and walked straight up to it; this time pretty sure that he had got what he wanted. When he was almost there the Scoutie rose with a derisive chuckle from a black-backed Gull's nest, where, as he had been slow in coming, she had whiled away the time by sucking one of the eggs.

But for Skuas, as for prouder potentates, "there is no armour against fate." We brought home, as a remembrance of an enjoyable day, the tail of one which had bowed to higher power and been eaten by a Hawk.

The Great Skua, which is three times the size of "Richardson's," breeds still on one or two of the northern islands, and on Foula, but is every year becoming scarcer. We did not see it ourselves in the Shetlands, but in the autumn, a year or two before, had fine opportunities of studying its habits, and realising the appropriateness of its scientific name, *Lestris catarrhactes*—the pirate who makes his descents with the dash of a waterfall—when, in company with three yachts and humbler sea-fowl innumerable, one of these magnificent birds was driven by stress of weather outside to run for shelter to Loch Broom.

The day after our visit to Noss, when on the point of No-Ness, fifteen miles or so south, we were taken to see a perforated rock, like a double arch of a submerged cathedral, which for many years had been the nesting-place of a pair of the Great Black-backed Gulls, worse tyrants, if possible, than even the Skua. The "Great Black-back" is a solitary bird, bearing, "like the Turk, no brother near his throne," dreaded and shunned by other birds, whose eggs and young he destroys.

Macaulay, minister of Ardnamurchan, and historian of St. Kilda, a great uncle of the historian of the larger neighbouring islands, writing in 1758, says:—

"It is hardly possible to express the hatred with which the otherwise good-natured St. Kildans pursue these Gulls. If one happens to mention them it throws their whole blood into a ferment. If caught, they outvie one another in torturing this imp of hell to death. Such is the emphatical language in which they express action so grateful to their vindictive spirit. They pluck out his eyes, sew his wings together, and send him adrift. . . . They extract the meat out of the shell of his egg and leave that quite empty in the nest. The Gull sits upon it till she pines away."

From the cliff where we lay down to watch them we could see three little birds—offspring of the feathered Cain—just out of the egg, lying on the short heather which covered the top of the rock, while the parent birds, whose consciences, perhaps, made cowards of them, hung near enough to watch us, but far enough off to have been well out of gunshot if we had had any murderous designs, which was not the case.

On the following morning, with a spanking breeze behind us, we sailed across to Mousa. The castle, which stands only a few yards from the shore, on the west side of the island, is probably the oldest building in the British Islands in anything like a complete state, and is of almost startling interest.

Ruins of squat round towers, known as "brochs," built of stone without mortar—the connecting link, according to Sir Walter Scott, between a fox's lair in a cairn and a human habitation—of which nothing is known, excepting, perhaps, that when the Vikings made their first descents a thousand years or more ago they found them standing and took possession of them—are scattered plentifully on the cliffs of the mainlands and islands of the North of Scotland.

The Broch of Mousa is the only one in existence which still stands, in all essential particulars, as in all probability it stood when originally occupied. It is a circle of stone wall about forty feet high, shaped like a chess castle with the battlemented top cut off The outside diameter is about fifty feet at the base and thirty-eight or forty feet at the top. It is bearded on the outside with a venerable growth of grey lichen, and tapers gradually from the bottom, until, within a few feet of the top, it slightly widens

THE BROCH OF MOUSA—SECTION.

again, so that the actual top almost imperceptibly overhangs.

It is not easy, without going into too much detail, to give an intelligible description of a building so completely at variance with every modern idea; but the very rough sketches given above and on the next page—

the one a ground plan, the other a section—may make it easier to understand the internal arrangements.

On the ground floor are three roomy domed chambers (marked in the ground plan B, and in the section G), built in the ordinary style of beehive-houses, with stones overlapping inwards to a point. The chamber that we examined, which was, we were

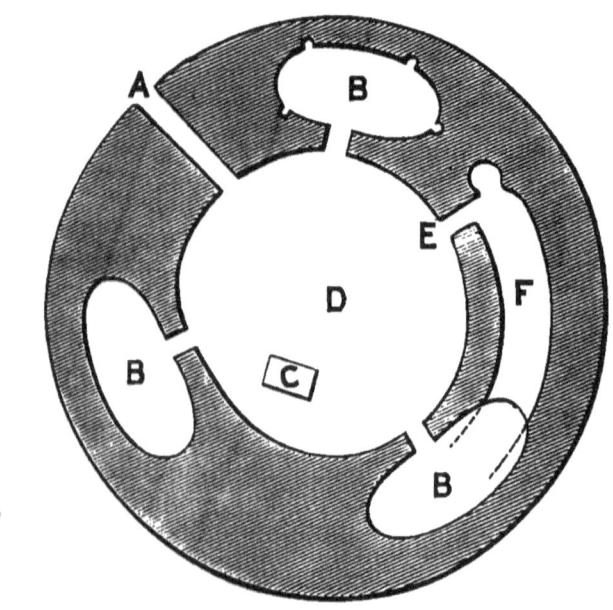

THE BROCH OF MOUSA—GROUND PLAN.

told, a fair type of all, was fourteen feet long, five feet six inches wide, and in the middle over nine feet high. It was entered from the court, in which was a square well or tank (C), by a low door (L), little more than three feet high, and less than two and a half wide, and ventilated, not lighted, by a small square window (M) immediately over the door. On one side of the chamber, there was a long narrow

Birds'-Nesting Season.

projecting stone shelf a foot or two from the ground, and built into the thickness of the walls were four or five neatly-shaped " ambries," or store cupboards.

If the fair mother of Harold of Orkney, a second Helen of Troy, who, in the days of King Stephen, was carried off to Mousa by Harold Erland, was a lady of as much taste as beauty, she may, with the help of a few gay coloured bullock skins as sofa covers and curtains to keep off the draughts, have made herself very comfortable as things went in those days during the long siege which the castle stood before her marriage with the turbulent lover who had compromised her.

The wall to the height of the top of the chamber dome is (excepting the chamber spaces) a solid heap of stone, between fifteen and sixteen feet thick. A little above the level of the tops of the chamber domes the wall divides, and thence to the top of the castle is built double in two concentric circles. In the hollow between the two walls a staircase (F), or rough stone path, not unlike the paved gradient by which the horses reach the stables over the coach-houses of Marlborough House, entered from the court by a door (E and I), leads up to the top of the castle, and six horizontal galleries (H) run round the building, lighted by holes opening inwards (K). Each gallery ends abruptly a few feet from the stairs, and all are so arranged that no one could reach the top of the tower without stooping and exposing himself to a knock on the head from an unseen enemy at each successive stage. The only break in the outside wall is at (A) the entrance to the courtyard.

Unless, as is not impossible, the walls have been nipped by settlements, the Picts, or whoever else they may have been who first designed the castle and burrowed their dwellings in the green slope behind it, must have been a race much smaller than the better fed man of the nineteenth century. It was only at some risk of being set fast, like a too keen fox-terrier in a rabbit's hole, that a pair of shoulders of not much more than average breadth could be pushed a little way through some of the most roomy of the galleries.

They, poor people, and the Norsemen who robbed and exterminated them, have their successors now in the Rock Pigeons, who have made a dovecote of the castle, and the Falcons who prey upon them. In the enclosed court lay the clean picked bones and feathers of a Pigeon killed a day or two before our visit, and just inside the entrance to the staircase, in a hollow under a stone, a naked nestling lay dead beside a cold egg, in which was another young bird, which, when the mother left the nest to return no more, must have been within an hour or two of hatching. In the corner of one of the chambers crouched a pair of young birds almost ready to fly. As we climbed the stairs a second pair, full grown but still uneducated, fluttered before us, and as we came out on the top of the tower, a Peregrine poised himself for a moment, and circling once or twice without any visible movement of the wing, sailed off magnificently to the north-west, probably to join his mate on the Noup of Noss.

There is a herd of Shetland ponies on Mousa. They are kept for breeding purposes only, and lead a life as free as the mustangs of Mayne Reid's stories. All the mares, with a single exception, had, when we saw them, foals beside them, and were kept well in hand by their shaggy lord and master, who, when he thought we had looked long enough, gave the order to move off, and when one mare lingered behind the rest with a tiny foal not many days old, which skipped about like a lamb, and looked scarcely bigger, he cantered down and at once drove her up. The stallions' place as they move is last in the herd. The standard height for a Shetland pony is 40 inches, and the present value of a fairly good one not taller, from £15 to £20. Many of them, poor little creatures, leave their island to spend the rest of their lives in coal-mines; but there has lately been a considerable demand from America, and many now go there.

On leaving the castle we made a circuit to the south-east, gathering a few common eggs for cooking, and crossing a beautiful bay of shining sand composed entirely of powdered shells of every shade of white, pink, yellow, and blue.

The cliffs here are very irregular. In places little caves, running in some way, have been bored by the waves and loose rocks, and as we walked near the edge, from underneath our feet came uncanny sounds —whisperings of young Starlings and underground rumblings and boomings of the sea, as if Trolls and imprisoned giants still lingered on the island.

Once a Lark rose close by us from a nest so well concealed that we looked without finding it, until, as if by magic, four Kingcups—the wide opened orange mouths of as many little birds just hatched—with chins touching, and necks stretched out till they looked a single stalk, shot up from the short heather and burst into full blossom at our feet. A few yards further on we picked up a baby Lapwing, which was doing its best to hide under a tussock of grass. The shepherds say that young Ringed Plovers are even more wide awake, and that a chick just out of the egg, when hard pressed, will grasp a dead leaf between its legs, and, rolling on to its back, lie completely hidden under it until the danger is past. But it was getting late and the wind was against us, and pleasantly as another hour or two might have been passed on Mousa, we were obliged to tear ourselves away. It was not until we had tacked six times that we found ourselves on shore again at Sandwick, in time and with appetites for an excellent dinner.

The teeming bird life of the Shetlands is confined, during the breeding season, mainly to the coast line. In the drive of five-and-twenty miles from Lerwick to Sumburgh, the last half of which we took the morning after our visit to Mousa, and in our walks across the Island to and from Scalloway, we were struck with the comparative scarceness of birds when out of sight of the sea.

Wherever there were buildings, the ubiquitous House Sparrow was, of course, to be seen, but not in anything like the numbers it is usually found elsewhere, and once, not far from Sandwick, we certainly thought we saw a pair of Tree Sparrows. But a treeless island is scarcely the place to look for a bird so named, and as we afterwards failed to find any mention of it in Dr. Saxby's "Birds of Shetland," and were too modest to suppose that it had been reserved for us, in a week's visit, to make an addition to his list, we were obliged to conclude that, to our eyes, more accustomed to the smoky colour tones of London, the clean head feathers of a spick and span House Sparrow in wedding garments had seemed the chocolate cap of the smaller and rarer bird.

The small birds we noticed oftenest inland were Mountain Linnets or "Twites," which, though scarce farther south, here take the place of the common

Linnet, which is seldom or never seen in Shetland. The two birds are very much alike, the only points of difference of any importance being that the beak, which in the common Linnet is a blue-black, is yellow in the Twite, and that the pink, which is a conspicuous feature in the summer plumage of most of the family, instead of appearing, as it does in the Linnet, on the head and breast, shows itself less strongly in the Twite on the back near the tail.

Every now and then what we took to be a Raven flew over, high up, or a Plover rose and wheeled round us, the hen bird waiting, as in Shakespeare's day, till "far from her nest," to cry "away," and trying to mislead us by doubling signs of anxiety, probably, as we walked away from her treasures.

We noticed a few Larks and Pippets, and occasionally a pair of Wheatears, who, like other visitors from the south, evidently appreciate the softness of Shetland wool, and were usually to be seen busily collecting it for a nest hidden in some snug corner under a rock not far off.

The value of Shetland wool in eyes other than those of breeding birds varies with the colour, the shade most highly prized being a cinnamon brown, known as Murad, not unlike the colour of the back of a ruddy Sheldrake—for which as much as half-a-crown a pound is often given before it is spun.

We felt a little as Moses must have felt on Pisgah, when, on reaching the top of the last hill before dropping down to Sumburgh, we saw across the Roost the outlines of Fair Island, looking, in the clear shining after the rain, not half its real distance and tantalisingly near.

Calm though the water had looked from the top of the hill, it was too rough to allow us, as we had hoped, to explore "the Head" from the sea, or to attempt anything with a small boat in the open.

But between Sumburgh and the towering precipice of Fitful Head, at the entrance of Queendale Bay, there are two islands well worth a visit. By the kindness of the owner, Mr. Bruce, of Sumburgh, a boat had been sent for us overland on a cart to a sheltered corner, and after a row of half an hour, during which we were objects of great interest to a party of seals, who popped up their heads and lifted themselves breast high to stare at us, we managed to reach them with clothes comparatively dry.

We had expected to find on the Lady Holm a fine show of Gulls' eggs and one or two nests at least of the Eider Duck, of which a few pairs commonly breed there. But, unfortunately, we were a day too late, a boatload of boys having, as we afterwards learned, effected a landing the night before, and made a clean sweep of every egg that could be carried off. Parties of Gulls stood in disconsolate attitudes by empty nests in every direction, and Oyster-catchers and smaller waders rose piping in a half-hearted manner to tell the tale that they had nothing left to lose.

The only birds which seemed thoroughly contented and happy were the black Guillemots, whose nests are very hard to find, and often, when found, as hard to get at. They rode peacefully at anchor in parties of ten or a dozen in every little bay, rising and falling with the swell of the water, one or other, every now and then, rousing himself just enough to lift a carmine leg to scratch the back of his head, or peck at some little fish or other tempting morsel which happened to float within easy reach.

But the interest of the islands is not dependent

only on birds' nests. On the smaller of the two are still to be seen the traces of a little chapel, probably, like many others in sites as lonely and picturesque, first built as a retiring place by some long-forgotten Culdee who has left behind him the only record of a saintly life in the name—" Cross Holm "—which the rock still bears. The beauty of the larger " Lady Holm," on the west side a heap of huge bare boulders, tossed up by the Atlantic rollers, which in winter gales half sweep the island, on the other side a level sward of sea-pinks, would alone have paid us well for our splashed jackets. But "Lady Holm" has a special interest of quite another kind.

The Shetland Islands seem, in the days when the world was being fitted up for human habitation, to have been used by Nature as an experimenting ground, and raised and submerged and raised again, heated and allowed to cool on no intelligible principle, scoured with ice, sometimes this way, sometimes that, until, as it now exists, it is hopeless for any but the most specialised of specialists to pretend to understand anything of the general geology of the group.

But a few things seem to come out fairly clearly. One of these is that once upon a time the promontory of Fitful Head must have been much bigger than it is now, and that, during this time, it was violently cracked, and that through the crack melted rock from very far below boiled up to the surface and hardened there.

Lady Holm seems to be a part of the original promontory as it existed at the time of the crack, which held its own when Queendale Bay was scooped out. The line of the intruded rock which crosses Fitful Head, if prolonged, runs through it, and accordingly we find a little island built up, in two clearly divided and nearly equal halves, of widely differing rocks. The wild western side is granite, and the gentle, richly flowered eastern slopes are sandstone.

Three or four miles from Lerwick the south road divides ; one branch zigzags along the coast towards Fitful Head, the other strikes across the island to Scalloway. On our return from Sumburgh we left the carriage at the parting of the ways, and sending it on to Lerwick with our baggage, walked across to Scalloway. The road undulates between hills covered with peat. Though it is in a way picturesque, there is nothing very striking to be seen, until, on the top of the last rise, the little port, with its beautiful land-locked harbour, lakes, and ruin, with the grand outlines of the hills of Foula in the distance, comes suddenly into view. The castle, which is unroofed, is of the common Highland sixteenth century type—a tall, square building, with high pitched gables, oriel windows, and round corner turrets. There is a coat-of-arms over the doorway, and conspicuous on the highest point of the western gable the iron ring from which tradition says that the founder, Patrick Stuart, of infamous memory, was in the habit of hanging neighbours who disagreed with him as to the fair price for their estates.

It is not difficult, without any greater mental effort than is involved in looking up the index references in the published Registers of the Privy Council of Scotland, to draw for oneself a fairly distinct picture of the man and his times.

Patrick was a grandson of James V. Robert Stuart, his father, had been Prior of Holyrood, but exchanged his priory with Adam Bothwell, the first Protestant Bishop of the See, for the bishopric or temporalities of Orkney.

The union of Robert Stuart's father and mother—the latter a young lady of high degree, who afterwards married a Bruce—had not been blessed by clergy; and, perhaps, on this account, the new bishop seems to have considered himself absolved from any oppressive obligations to the Church. He persuaded the king to make the bishopric an earldom, and at once set to work in his own fashion to increase his estates in Orkney and Shetland. If Church matters were managed now in Scotland as they were then, Dr. Cameron might be pretty sure of a majority when next he raises the question of Disestablishment.

Robert, the father, had chastised with whips. Patrick, the son, was to chastise with scorpions. In the Council Registers of the last few years of the sixteenth, and first few years of the seventeenth centuries are entered constant complaints from poor Orkney men and Zetlanders of oppression, such as had never before been "hard of in ony reformed cuntrey subject to ane christiane prince."

Earl Patrick steals Sir Andrew Balfour's sheep, cows, butter, and seed corn, and "refts from him and his puir tennentes, twenty-nine whales, which at grite charges and expenses," they had driven on shore on Sir Andrew's own land. He besieges and takes away Sir Patrick Bellenden "(he being 72, in a wand bed), and delivers his hous to Keipers, and all because he would not despone his londs to him," and so on until "no man of rent or purse might enjoy his property without his speciale favour, and that same dear bought, filchit and forgit faults being

G

so devisit against many of them that they were compellit by imprisonment and small rewaird to resign their heritable titles to him . . . . gif not life and all besides."

It is not difficult to understand why, after most entries of the kind, we read, "Wanting probation the earl is assoilized," as at least ten times in a single volume of the Register appear such entries as the following:—

"*Sederunt*, Cancellarius, Orknay Thesaurius, collector, &c."

"*Sederunt*, presente Rege, Lennox, cancellarius, Angus, Orkney, Mar, &c."

But Lord Orkney trod once too often on the toes of his Royal cousin, and in 1613 Lord Carew,[*] writing to give his dear friend, Sir Thomas Roe, ambassador to the Great Mogul, the last gossip of the London season—that Sir Moyle Finch is dead, leaving the richest widow in England; that Lord Berkeley and Lord Fitzwalter have married the two pretty daughters of Sir M. Stanhope; that a ship fitted with provisions for nine months (the forerunner by 200 years of Sir John Franklyn's ill-fated expedition) is just starting to find a North-West Passage; and that there is much talk at Court of the "rising fortune at Court of a young gentleman 'of good parts'"; a Mr. Villiers, &c., is able to fill a corner in his letter with the news that "the Erle of Orkeney in Scotland is beheaded and his lands and honnour escheated to the Kinge."

As we left behind us the beautiful scene of so many iniquities, a Raven, big and hoarse enough to have been a survivor from Patrick's day, when

[*] "Letters of Lord Carew." Published by the Camden Society.

Ravens' food was cheap in Scalloway, flew close over us, croaking an appropriate good-bye.

It was a farewell to the Shetlands, as well as to the castle.

On reaching Lerwick we found at the quay a steamer which was to sail that night with a cargo of fish and cattle, direct for Aberdeen, and as the weather was still broken, and there was little more that we could see, we put our things on board at once, and three days later had crossed the Forth Bridge, the first day it was opened for general traffic, and were in London again.

For those of us, especially whose place in the procession of the generations happens just now to be among the workshops on the table-land of middle life, it is wholesome to be reminded every now and then that time is a created thing, and life possible without its limitations.

It is a pleasant reminder of the kind to look back on a holiday-trip into which the impressions of twelve months seem to have been crowded, and to know that while one has been away from home the sun has only risen and set on as many days.

The Last English Home of the Bearded Tit.

"When Ducks by scores travers'd the Fens,
Coots, Didappers, Rails, Water-hens."
"*Antiquarian Hall*,"
The Fen-Man.

IN the memoir of the Geological Survey of the country round Cromer is a rough sketch-map of the outline of the north-west corner of Europe as in all probability it existed at the Newer Pliocene period, in the far-off days when the primitive vegetation and monstrous creatures of a still earlier world were slowly giving place to plants and animals of "more of the recent" types.

A great river, since dwindled to the insignificant Rhine, with its mushroom castles and ruins, swept through fir woods and swamps to an estuary hemmed in to the westward by a coast-line unbroken, excepting here and there by a tributary stream, to John o' Groat's, rolling down in its sluggish current stumps of trees and bones of elephants and bears and beavers, to be washed long ages afterwards from the "Forest Beds" of Sheringham and Runton.

The swamps through which the old estuary once cut its way lie buried now in places a hundred feet and more deep beneath Norfolk turnip fields and pheasant coverts.

The fens of the Great Level, which, before Dutch drainers and dyke-builders had reclaimed the second Holland, were perhaps their nearest counterpart in the England of human times, are scarcely less things of the past. The marsh devils, which, until St. Bartholomew interfered and drove them off with a cat-o'-nine-tails, held open court there, and, as Matthew of Paris tells in his Greater Chronicle, came out in troops to maltreat the few hardy Christian settlers who, like St. Guthlac, as penance for past wild lives, sought holy retirement there—dragging them, bound, from their cells, and ducking them mercilessly in the black mud, "cœnosis in laticibus atræ paludis"—now cower invisible in the ditches, or sneak out as agues, to be ignominiously exorcised with quinine. Hares and Partridges have taken the place of Spoonbills and Bitterns, and Ruffs and Reeves; and, where a few years ago wild Geese swam, ponderous Shire cart-colts gallop, scarcely leaving in summer a hoof-mark on solid ground.

The old order almost everywhere has changed and given place to new. But there is a corner left—the district of the Broads of Norfolk—where one may still see with natural eyes what the world in those parts must have looked like in days before the chalk dam which connected England once with the mainland was—happily for Englishmen of these days—

broken through, snapped by a sudden earthquake, or slowly mined by countless generations of boring shellfish, until it gave way under the weight of the accumulating waters of the estuary, choked to the north by advancing ice, or tilted westward by some submarine upheaval. There, with a very small stretch of imagination, one may still hear mastodons crashing through the reed-beds, and British hippopotami splashing and blowing in the pools; and, as every now and then an incautious footstep breaks through the raft-like upper crust of soil, and imprisoned gases bubble up, one may, without any stretch of imagination, smell the foul stenches of Pliocene days.

The climate in those days, geologists tell us, judging by the fossil plants of the time, must—before the country was wrapped in ice—have been much what it is in Norfolk now. "If the various sections of the upper fresh-water beds are examined, we find," writes Mr. Clement Reid, who surveyed the country round Cromer, where the Forest Beds are most exposed, "that all appear to have been formed in large shallow lakes like the present Broads, or in sluggish streams connected with them."

Three considerable rivers, the Bure, the Waveney, and the Yare, after meandering through level meadows and marshes—none of the three, according to Sir John Hawkshaw's estimate with a fall of more than two inches in the mile—join and meet the full strength of the tide in Breydon Water.

The outflow is checked and the volume of the streams, finding no other way to dispose of itself, has spread out into side-waters and back-waters, wherever the law of levels, the only law to which it owns allegiance, has admitted a right of way.

The result is a triangle of some fifteen or twenty thousand acres or more in which, as in the abyss through which Satan winged his way in search of the newly created world,

> "Where hot, cold, moist, and dry, four champions fierce,
> Strove for the mast'ry,"

land and water hold divided empire. In places the water seems at the first glance to be carrying all before it. Broad sheets (some of them a hundred acres or more) spread almost unbroken surfaces over unfathomable depths of mud. But the encircling rings of rushes, dwarf alders, and other multitudinous marsh plants, creep in insidiously, each generation growing rank and dying to make soil on which the next may find a footing for another step inwards.

The water revenges the encroachment by flooding the land wherever it finds a chance, and undermining when it cannot overflow, till it is impossible to say where the one begins and the other ends. One walks almost dry-shod across what had seemed a dangerous pool, and the next moment sinks over one's fishing stockings in what anywhere else would have been dry land. The confusion of ideas as to the relative solidity of earth and water which results from an hour or two spent in exploring a soft "Broad" marsh is not lessened as one sees the huge brown sail of a "Wherry"—the craft which is said to go closer to the wind than any other afloat—moving straight up to one, to pass by at eight or nine miles an hour, sailing to all appearance on dry ground. The navigable channels are most of them natural cuttings in the dead level of the marsh, invisible at a very few yards' distance.

The name of the long pole, which is one of the

most important parts of the equipment of the Norfolk wherry—the " Quant "—is, by the bye, a memorial of the days of Roman occupation. It was with a quant, spelt a little differently in Virgil's day, that Sergestus in the immortal boat race tried to shove off his galley when he had cut his corner too finely and run aground; and with a quant that Charon ferried his passengers across the Styx :

> " Ipse ratem conto subigit velisque ministrat."

The entire district is unlike anything else in England, and, apart from its power of recalling the past, has an exceptional interest of its own for naturalists. It is the paradise of shy creatures of all sorts, birds especially, which love mud, or water, or reeds ; and has been the last settled English home of more than one rare species. Their number, in spite of the keener interest taken of late years by landowners in bird preservation, steadily decreases.

The Avocet, with its spindle shanks and beak turned up like a shoemaker's awl, which not very long ago bred so freely in the salt marshes that "poor people made puddings and pancakes" with their eggs, is now the rarest accidental visitor. The Bittern, comparatively lately a regular breeder there, no longer "guards his nest" among the sedges and reeds ; and Ruffs and Reeves are as rare as they once were common. But there is—or at least till last year was—one little bird which, driven from every other part of England, has made the Broads his own peculiar property, and himself thoroughly at home there. Hardy and modest in his wants, the Bearded Tit has been essentially a home-staying bird. His ancestors seem to have elected, generations ago, that, whatever the advantage of a winter in Algeria, the disadvantages were greater, and that, on the whole, it was better to face the evils that they knew than fly to others that they knew not of.

The "developments" of the family ever since the decision was made have been in a direction to fit them for a quiet life among the reed-beds. Other birds, smaller even than they, whose forefathers were of a different opinion, have wings now so perfected that, when soft animal food fails in England, they think nothing of a flight of a few hundred miles to a sunnier spot where fat insects may still be found.

The Bearded Tit, with his little round wings and the heavy canvas of his long tail, cannot do what they can. But he can do what they cannot, and make the most of what is to be got in the way of food at home.

In the swampy grounds from which his reed-beds grow are quantities of very small snails. Some early ancestor, feeling the pinch of hunger, ventured experimentally to pick one up and ate it, and finding out the sustaining qualities of the rich inside meat, brought up his young ones to eat them too, and make light of the aches which a sharp-edged, hard shell swallowed whole must have caused in a delicately-coated stomach.

They, in their turn, brought up their young on the same Spartan system, and now—unlike other Tits which have most, if not all, of them tender insides, suitable enough for digesting soft insects, but unfit to do justice to anything harder than a seed well steeped in gastric juice—the Bearded Tit finds himself the possessor of an honest, sturdy gizzard, which can grind up without the least inconvenience to the owner any number of the shells of the snails which are its chief delicacy. As many as twenty little snail-shells have been taken from the crop of one Bearded Tit.

We wonder now why good people should have been so much alarmed as once they were at the doctrines of "development." It is the teaching of the Parable of the Talents extended from the spiritual to the physical world—powers neglected or abused withdrawn, others well used increased.

The shape and colour of the Bearded Tit are as specially adapted as is its stomach to the peculiarities of its surroundings.

Visitors to the Broads in midsummer who may have caught glimpses of the bird, showing itself for a minute or two at a time, a conspicuous object against the green of the young rushes, may find it difficult to realise that the Bearded Tit is, when invisibility is of most importance to it, protected by colour and form scarcely less perfectly for all practical purposes than are leaf-insects, or stick-caterpillars, or the wonderful creatures described by Professor Drummond in his "Tropical Africa."

But such is the case. The eggs are laid about the middle or end of April, when the tall reeds, among which the nest is built an inch or two from the ground, are ripe for cutting.

The prevailing tints of the entire district—land, water, and sky—are then the cinnamons, straw colours, and pale blue greys, miraculously reproduced in the feathers of the bird, which might pass for the emancipated spirit of the dead reeds of last summer. The long tail, with its pointed end, hangs down as its owner comes in sight for a moment to look about him, the counterfeit presentment of a faded frond of the stalk he grips, one foot below the other.

The Hoopoes, as the legend goes, wear their

crown of feathers in memory of the day when their ancestors saw King Solomon almost fainting under a sudden burst of noonday sunshine, and sheltered his royal head with a parasol of overlapping wings.

It may be as a mark of approval of the manliness with which he faces winter on the Broad, when Snipe and other birds have been driven off by the cold, that the Bearded Tit now wears the long silky black moustache — his own peculiar adornment — which hangs from each side of the beak.

As in the nobler species, the moustache is noticed only in the males. There is a prolongation of the cheek feathers of the female also, but not the same contrast of colours.

For all ordinary winters the Bearded Tit is well provided. But, unhappily, last winter—the longest on record since the days of Lorna Doone—was not an ordinary one.

Fifty-nine days of consecutive, almost sunless, frost were recorded in London, and in parts of the Broads the weather was even more severe. The snails for weeks and months must have been glued fast to the ground or rush-stalks—tantalisingly in sight for much of the time, as there was no great quantity of snow, but as much out of reach of a small beak as flies in amber. The birds when most in need of a warming meat-diet were driven to depend almost entirely on such dry ship-biscuits as the seeds of reeds, without even water, excepting here and there in the running streams, to wash it down, and have suffered terribly in consequence.

It was on one of the bright mornings towards the end of April last, when, in spite of a wind still nailed in the east, a warm sun and such spring sounds as the call of the Nuthatch, a pair of whom had from daybreak been carrying on a lively conversation over an unfinished nest in a box in the garden, encouraged the hope that the return of the glacial epoch might not after all be so near as for the last six months had seemed probable, we found ourselves, after an early breakfast and drive of fourteen miles, landing from a boat on the edge of a marsh skirting a Broad. The marsh is strictly preserved, and on it, as lately as last summer, Bearded Tits were plentiful. We had come in the full expectation of seeing both birds and nests, and were, if anything, rather encouraged than otherwise when the keeper—in the pessimistic tone common to men of his order when conscious that there is an unusually good head of game in front of the guns—told us that, though there was a nice lot of reeds uncut, he "doubted" we should not find any Tits, as to the best of his belief there was not one of them left in the place.

But before an enjoyable day was over his words had acquired a different meaning. We tramped the marsh, which teamed with other bird life, backwards and forwards. Twice we flushed a Mallard from a nest well filled with eggs. One nest, with a clutch of ten, was downed almost as thickly as an Eider Duck's, with a well trampled path like a miniature sheep-walk leading from it to the water's edge. From behind a stook of reed-sheafs we watched for ten minutes a pair of Teal playing together—unobserved, as they supposed—in a rushy pond close by.

Shovellers, with fantastic colouring and great flat beaks out of all proportion to the size of the bird, rose more than once within a few yards of us, and after circling once or twice, pitched again not far off.

Tired-looking Swallows sat disconsolately in parties

of five or six on bushes, or rose to skim over the water in a half-hearted way, and light again.

A pair of Redshanks crossed us once or twice, flying in line, one just behind the other, whistling loudly as they flew. Cuckoos called, and overhead Snipe poised themselves, drumming and bleating, and dropped like stones as they neared the ground. In the nest of one of them we saw a beautiful instance of "protective colouring," the marvel of which never loses its freshness.

The keeper the day before our visit had found the nest, and for our benefit had marked the spot. It was in a line between two bushes, within a half a dozen yards of one which stood alone and unmistakable on flat ground, with nothing on it bigger than a few short sprits which could hide the nest. As we neared the spot, the bird, to show there could be no mistake in the mark, rose close by us.

For more than a quarter of an hour we looked—three pairs of eyes, one pair the keeper's—crossing and recrossing every foot of the ground, and were giving up the search as hopeless, thinking that a Crow perhaps had hunted the marsh in the early morning before us, when in the middle of a tussock of sprits at our feet we saw a Maltese cross of very green eggs, mottled irregularly with brownish-red, exactly imitating the bed of deep moss from which the sprits grew.

The colour of Snipes' and many other eggs is very volatile, and no one who has only seen them "blown" in a cabinet can quite realise their beauty when seen in the nest, fresh-laid and untouched.

At intervals of our tramp on shore we took the boat rowing across corners of the Broad, or pushing our way through ditches or narrow twisting channels. We saw Coots' nests in plenty, and one unfinished nest of the Great Crested Grebe—the one rare bird which has made some return for the trouble taken of late years for its preservation by becoming more common. A floating mass of weeds, fished up, wringing wet, from the bottom of the water, looks a hopeless nest for a bird to hatch her eggs in; but, like a damp haystack, it generates very considerable heat.

"In a Grebe's nest," writes Mr. Southwell in the third volume of "Stevenson's Birds of Norfolk," just published, "in which were three eggs and a newly hatched young one, the thermometer rose to 73°, showing that the nest, so far from being the cold and uncomfortable structure by some supposed, was a real hotbed. On inserting the thermometer into a beautifully neat and dry Coot's nest, which the bird had just left, I found the temperature to be 61°. The day was wet and cheerless, and the maximum reading of the thermometer in the shade was 58°."

We saw through our glasses several Crested Grebes playing on the Broad. Oddly enough, the common Little Grebe—the "Dabchick"—is less plentiful in Norfolk than it is in St. James's Park, where last year as many as six pairs, all wild birds, nested and brought off their broods.

For six or seven pleasant hours we hunted marsh and Broad with eyes and ears open. But not once did we catch sight of a feather, nor once hear the silvery "ping" of the note of the Bearded Tit.

It was, of course, one corner only of a wide district over the whole of which the bird has been well known that we had explored. There are other Broads and marshes where local circumstances may have tempered the killing wind. There, while we looked for them in vain, busy parents may have been working hard from morning till night to cater for the wants of hungry families safely hidden in daily thickening growths of bog-flowers and grasses, and another year the deserted reed-beds we visited may be re-peopled.

But as we drove home the conviction forced itself more and more strongly upon us, that from one at least of its most favoured haunts the Bearded Tit has disappeared, and that it is not improbable that very soon—perhaps before this year is over—naturalists may be telling the sad story of the extinction of one more English bird.

WELL! I AM BLOWED!!

ST. KILDA from without.

> "Prithee, do not turn me about;
> my stomach is not constant."—
> *The Tempest.*

"MOTHER," said a little boy, whose knowledge of life was mainly confined to South Kensington, at the end of a glowing description of Eden before the fall:—

"A happy rural seat of varied hue—"

"Mother, I suppose it is all built over now?"

The little fellow's remark fairly represents the state of mind of the average country-bred Londoner towards the end of June. This world by then seems all bricks and mortar; and, if in her better moments the uneasy soul does ever try to look beyond present crowded streets, and expatiate in a world to come, the chances are that, even if the attendant body is not first knocked down by a passing hansom cab, she is met and frightened back to earth by bewildering thoughts of the myriads of past generations of every shade from ivory to ebony who have lived and passed on somewhere.

At such times it is wholesome to remember, even if with little hope of ever seeing it, that there is still in British waters one island at least where the population never increases, where the post comes in only once or twice a year, and where the conditions of human existence are much as they were a hundred if not a thousand years ago.

The hills and valleys which gives so much of its picturesqueness to the North West of Scotland do not end at the coast line, but are continued, as soundings show, under the sea some eighty or a hundred miles beyond the Hebrides. At the extreme western edge of this underwater, mountainous tract—which contains, perhaps the best fishing grounds in Europe —just before the bottom settles down to ocean depths in the open Atlantic, is a small oval bank of shallow water, from which rises abruptly, like the peaks of a submerged mountain, a little cluster of precipitous islands.

Until a year or two ago, when summer excursion steamers began to visit it more regularly, St. Kilda, or, as the natives still prefer to call it, Hirta, the chief and only inhabited island of the group, was comparatively unknown.

Every now and then it emerged for nine days from its obscurity, when a corked bottle or toy ship, carrying a message on which life or death depended, was washed ashore somewhere on the opposite coast, announcing, as in 1877, that an Austrian ship had been lost on the rocks, and that, with the extra mouths of the rescued crew to be filled, provisions could not hold out long, or, as in 1885, that a storm had swept the Island and destroyed the crops. Once in its history St. Kilda has been honoured with a Parliamentary Debate to itself.

In 1869, when the first Sea-bird Preservation Bill was under discussion, the Duke of Northumberland rose in his place in the House of Lords to move the addition of a clause:—" The operation of this Act shall not extend to the Island of St. Kilda."

The amendment was opposed by a second Duke, who, in advance of his times, argued that where there was no policeman to enforce an Act it was unnecessary to enact that it was not to be enforced. A third Duke replied (no legislator of humbler degree took part in the debate), and,—the House very properly hesitating to give its sanction to the shocking morals propounded by the noble objector,— the clause was accepted without a division, and has since been embodied in subsequent Acts. Parliament has been before now likened to the elephant's trunk— which can pick up a pin or uproot a forest tree. But it would be difficult to find another instance in which the wants of a population of less than eighty* all told, have been provided for by special legislation.

There is a tradition that the sixty or seventy miles over which the Atlantic now rolls, between St. Kilda and the nearest point of the Hebrides, was once bridged by an isthmus, across which the hounds of a great huntress, " The Greatly-Savage, Soft-skinned, Red-haired Muiream," ran deer from Conagher to the Butt of Lewis.

The lady—whose fame Captain Thomas, R.N., in an interesting paper published in the Proceedings of

* The population of St. Kilda, as shewn in the Census returns, was

	MALES.	FEMALES.	TOTAL.
In 1861	33	45	78
1871	27	44	71
1881	33	44	77
1891	32	39	71

the Society of Antiquaries of Scotland, traces in the legends of both islands—like Samson in the Temple of Dagon, met her end in a way that did no discredit to her reputation, "the dead which she slew at her death being more than they which she slew in her life."

Her husband, the king's fisherman, was out in his boat, and had just caught a sturgeon, when a party of marauding Irish set upon him, stole his fish, and, on his mildly protesting, belaboured him with it within an inch of his life.

No one who has not seen such a club brought into active play is likely quite to realise what a formidable weapon a big fish in the hands of an angry man can be; nor, having seen it, as it was once the writer's fortune to do, is likely soon to forget the sight.

We were driving from the Baltic port of Abo, through pine woods and boulder-strewn mosses, cut up by streams and white lakes, and dotted here and there with cultivated patches and the log huts, commonly used in Finland for smoke-drying the crops, to a fishing-ground a hundred miles or so inland. A couple of hours of daylight had been wasted in an unsuccessful attempt to stalk some Cranes, which had flown over us with necks not bent backwards between the shoulders, as a Heron carries its neck, but stretched stiffly forwards. We had marked the birds down in an oat-field, a few hundred yards off the road, where, in spite of all our efforts to outwit them, the eyes of four sentries, who stood on duty, straight as gateposts, while the others fed, had proved too sharp for us.

The daylight had almost gone as, with several stages of the road to be driven still before us, we pulled up for fresh horses in the courtyard of a change-house, standing between lake and forest. The day's work was over. There was a noisy gathering of men and women and children in the yard, and the competition for the honour of driving us the next stage was keen.

A hulking fellow, in a sheepskin coat, in the quarrelsome stage of drunkenness, took possession of the box-seat of our carriage, and, refusing to give way, was seized by two or three others and violently ejected. As he rolled on the ground swearing, a boy, laughing and singing at the top of his voice, jumped up and slashed the ponies. The picture which we looked back upon, as we started at a gallop, followed by the baggage-cart, driven by a girl, is as fresh in the memory now as the day it was painted.

The sun had just set. Wreaths of mist were creeping up from the lake to the fir-trees which were massed in purple against a sky of transparent green and orange. In the middle of the inn-yard raged our friend, driving all before him as he hit right and left with a big salmon, snatched dripping from a pickling-tub, gripped with both hands by the tail.

Formidable enough as is a salmon as a club, it is nothing to a sturgeon, with its chains of diamond-shaped pyramids of outside bone—the last survivor, in European waters, of the mail-clad ganoids which struggled for existence among the crocodiles and flying monsters of a primitive world, and no wonder Muiream's poor husband, when carried into his master's presence to tell his story, was more dead than alive.

But he had not to wait long for his revenge.

The same night the lady crossed alone in a boat to Ireland, and, surprising the camp, killed the king's son and a hundred of his warriors, before her passion was stilled for ever by a stone dropped on to her head from above, as she stood, like FitzJames, with her back to a rock, defying the whole host of the Irish.

A curiously-shaped chambered beehive-hut, which has puzzled more than one antiquary, and a spring in St. Kilda, are still known as the "Dairy," and "Well of the Amazon."

The Soft-skinned Muiream is not the only peppery lady whose name holds a prominent place in the annals of St. Kilda.

In these days of woman's rights and newspapers, it is not easy to realise that only 150 years ago it could have been possible for a man, in a conspicuous public position, to deport a troublesome wife, and keep her in banishment for years, without apparently any inconvenient consequences to himself.

But this is what Lord Grange, one of the most distinguished lawyers of his day in Scotland, with the help of his infamous boon companion, the Lord Lovat, who, after Culloden, atoned for many abominations on Tower Hill, actually did.

Lady Grange had, by her own admission, a tongue in her head. "There is no person," she pathetically writes, "but has his faults," and, until adversity had broken her spirits, was not, perhaps, disposed to be as blind as a well-trained Eighteenth Century wife was expected to be to a husband's irregularities. They had agreed to separate; and she had taken lodgings in Edinburgh. What followed, is best told in her own words, written,* "with a bad pin," from St.

* The letter from which the extracts are taken is published at length in the "Proceedings of the Society of Antiquaries of Scotland."

Kilda, on the 20th January, 1738, to a cousin who, though she could scarcely then have known it, was, at the moment, Lord Advocate for Scotland.

"I lodged in Margaret M'Lean house and a little before twelve at night Mrs M'Lean being on the plot opened the door and there rush'd in to my room some servants of Lovals and his couson Roderick Macleod he is a writter to the Signet they threw me down upon the floor in a Barbarous manner I cri'd murther murther then they stopp'd my mouth I puled out the cloth and told Rod: Macleod I knew him their hard rude hands bleed and abassed my face all below my eyes they dung out some of my teeth and toere the cloth of my head and toere out some of my hair I wrestled and defend'd my self with my hands then Rod: order'd to tye down my hands and cover my face most pityfully there was no skin left on my face with a cloath and stopp'd my mouth again they had wrestl'd so long with me that it was all that I could breath, then they carry'd me down stairs as a corps at the stair-foot they had a Chair and Alexander Foster of Carsboony in the Chair who took me on his knee I made all the struggel I could but he held me fast in his arms My mouth being stopped I could not cry. All the linens about me were covered with blood."

A "Writer to the Signet" had been one of the party who carried her off, and it was in the house of an Advocate, "a little beyond Lithgow," that Lady Grange was first hidden. Thence a little later, without knowing where she was going, or what was to become of her,—at one time imprisoned in "a low room all the windows nailed up with thick boards and no light in the room .... left all aloan and two doors locked on me," at another time taken "naked" from bed "by force and put upon a horse where I fainted dead away"—she was carried to Huskre, a little island off Skye. The poor lady must finish her own story:—

"On the 14 of Jun: John Macleod and his Brother Nòrmand came with their galley to the Huskre for me they were very rud and hurt me sore. Oh alas much have I suffer'd often my skin mead black and blew, they took me to St. Kilda. John Macleod is call'd Stewart of the Island he left me in a few days, no body lives in it but the poor natives it is a viled neasty stinking poor Isle I was in great miserie in the Husker but I'm ten times worse and worse here." And yet she writes earlier in her long letter "He was my idol! He told me he loved me two years or he gott me and we lived twenty five years together few or non I thought so happy!!"

Lady Grange seems to have been about eight or nine years in St. Kilda before she was allowed to return, and then only to die soon afterwards in banishment, scarcely less complete, in Skye.

In many ways the solitary little group is of exceptional interest.

Antiquarians and students of men and manners may find subject for congenial speculation in the doubtful origin of the inhabitants and their four-horned sheep, and the identity of their patron Saint, unnamed in any calendar: and may read a curious illustration of the almost infinite gullibility of humanity in the story of the imposter Roderick—begun by Martin and finished by Macaulay—who, towards the end of the seventeenth century for six years or more, ruled supreme in the island, robbing the men wholesale, and debauching the women in the name of St. John the Baptist and the Virgin Mary.

St. Kilda from Without.

Advanced politicians may study, in a pure democracy, the working of a code of game laws as absolute and perhaps more rigidly enforced than in the most aristocratic country in Europe, even the Minister, who, until his retirement two years ago, was powerful enough to put a stop to whistling in the island, and to decree the observance of two sabbaths weekly, being as dependent for his Gannets and Fulmars on the good will of privileged families as any country parson on the favour of the squire for a brace of Pheasants.

For doctors to investigate there is the cnatan-nagall—the feverish epidemic cold, in some of its symptoms very like influenza—supposed to be peculiar to the island, described again and again for more than 200 years: laughed at by travellers from Dr. Johnson downwards, but none the less believed by the natives to follow almost invariably the arrival of strangers.

A more terrible mystery, still unexplained, is the "Eight days sickness," the infantile lock-jaw, which for years has carried off more than half the children born in St. Kilda, commonly, as the name given to the disease implies, on the eighth day after birth.

Miss McLeod, of Dunvegan, the good spirit of St. Kilda, thinking that unscientific nursing might be the true explanation, sent over not long ago a trained nurse. Great things were hoped from her services, and Mr. Connell, who soon after visited the island, wrote on his return that things "looked as if a rift in the cloud had really made its appearance, and that the high rate of infant mortality was to be a thing of the past." Recent returns, sad to say, have not justified the hope. In the first six months of 1891 three babies were born. Two of them died of the "Eight-day sickness."

But many as are the other interests touched, it is for ornithologists that St. Kilda reserves its chief fascination. In no other spot in British waters, not even excepting the marvellous "Noup-of-Noss" in Shetland, do sea-birds of many kind congregate in anything like the same numbers. It is the last recorded haunt of the Great Auk, and the breeding place of ninety-nine, at least, of every hundred Fulmars that nest within the haunts of the British Isles.

The Fulmar, which is the chief source of the islanders' wealth, supplying—to quote an old writer—"oil for the lamp, down for the bed, the most salubrious food, and the most efficacious ointment, and possessing a thousand other virtues," is a typical representative of its class, the Tubinarides, which,—varying in size from "Mother Carey's chickens," with bodies less than a Sparrow's, to the great wandering Albatross, with a stretch of wings of 16 or 17 feet,—are all formed on the same general lines, fitting them for a life spent almost entirely on the wing. The most marked characteristics of the family, next to the great development of wing muscles, are the nostrils, which, instead of being, as in most other birds, mere slits in the beak, take the form of prominent open tubes, through which the air—a free current of which is, it is easy to understand, essential for long untiring flight—passes to the inner air-vessels unchecked. In some of the tribes the tube-openings are at the side of the beak, ending, as in the case of the great white Albatross, with the upward curl of a cavalry officer's moustache. In others, the tubes run straight out like pistol barrels, single or double, resting and ending abruptly on the upper part of the beak.

What is given to the wing is taken from legs. A Fulmar in confinement, even if it can be induced to feed, commonly dies almost at once of cramp in the thighs.

When poor Lady Grange called St. Kilda a "stinking" isle, she used the word probably in a more literal sense than that attached to it in the ordinary schoolboy's vocabulary. The Fulmar is a living keg of strong, musty, scented oil, the smell of which is said to pervade the whole island and everything in and about it. The eggs, which in shape and colour are not unlike large, finely-grained, and very white and thin-shelled hen's eggs, retain the smell for years, strongly enough, when a drawer is opened containing one of them, to scent the entire room.

Three years running, with the special object of seeing the gathering of the Fulmars, we had perseveringly planned expeditions to St. Kilda. In 1889, enquiries, which failed in the end, as the locality is not popular with shipowners, were made for a steamer to take a party of congenial spirits for a few weeks' exploration. The next year—a humbler programme—berths had been engaged in the earliest tourist boat. But, as for two days before the time fixed for sailing a gale had been blowing, it seemed useless to rush off to Oban with the almost certainty of getting no farther. The third attempt was to be the lucky one, and everything looked well. The census was to be taken at the end of June, the height of the breeding season, and one of Her Majesty's ships, with a spare cabin, was told off for the trip.

On a lovely still evening—an eighth or tenth consecutive day of dead calm, after a night and day journey without a break—we left the train at Strome Ferry.

There was not a ripple in the land-locked bay where the two steamers—the one bound for the Skye ports, the other for Stornoway—lay with steam up, and, as we moved out, we could not have wished for fairer promise.

But alas!—

> "Not seldom evening in the West,
> Sinks smilingly forsworn."

There was an unnatural oiliness in the calm of the sea, and, as we cleared the successive headlands, and came more into open water, the masts began to sway slowly and steadily backwards and forwards, far out on each side. The cordage creaked and strained with the monotonous regularity of the snore of a heavy sleeper, and we were aware of an ominous ground swell, the reflex of tremendous waves somewhere. The surprise was less, but not the disappointment, when, on steaming into Stornoway Harbour late on Saturday night, the first news which greeted us was that the storm cone was hoisted.

Our ship, the *Jackal*, was timed to start about eight on Sunday evening, and by luncheon time we were on board.

The afternoon slipped pleasantly by, spent chiefly in watching one of the sights—familiar enough to those who occupy themselves on the great waters, but ever new to a landsman—which help one to realise the meaning of the extraordinary powers of reproduction with which fish are gifted. A shoal of small fry of some kind had found its way in, and was moving about the harbour with the usual escort, coming every now and then within a few yards of the ship's sides.

Great fish, glistening like silver, doubling in loops (heads and tails almost touching), were in the air together, six or seven at a time, mixing in wild confusion, and changing elements with screaming Gulls, Gannets, and Terns, which dropped like stones into the sea, while a Black Guillemot, keeping well clear of the ruck, dived in and out, his carmine legs flashing, and popped up time after time with a little fish in his beak, till the wonder was how he could possibly swallow another. In the height of the excitement a porpoise, puffing and wheezing like an asthmatic old gentleman in a hurry to catch a train, bustled up, passing within ten yards of us, to join the fun.

In the Ninth Report of the Fishery Board for Scotland, presented to Parliament in 1891, is a Report on the Comparative Fecundity of Fishes, by Dr. Wemys Fulton, which gives some figures which make it easier to understand how it is that, with so many enemies to contend with, any little fish live to grow up.

A single ling, which seems to be the most prolific of the many kinds reported upon, can, it has been proved, produce from 20,000,000 to 30,000,000 eggs in one season.

With such a record to head the list, such paltry clutches as 47,466 for a herring, and 806,459 for a haddock, seem too insignificant to be worth mentioning.

It was once calculated by Mr. James Wilson, the ornithologist, that the Solan Geese, breeding in the colonies of St. Kilda, alone must devour every year something like 214,000,000 fish.

At eight o'clock on Sunday night the cone was still up; but, as there was still but little wind, the anchor was weighed and we steamed out, but only to meet a white fog, which crept in from the Atlantic and drove us back to anchor again for another four and twenty hours.

At last, on Monday evening, the Butt of Lewis was rounded, and the ship's course shaped for St. Kilda. Eighteen hours later the anchor was dropped again in smooth water, this time in Loch Roag, a landlocked harbour to the north-west of Lewis, half a mile from a miniature Stone Henge which crowned a neighbouring slope. We had been within fifteen miles of our destination, and had been forced by a freshening gale to put back and run for shelter, and were not sorry when we reached it. All the crockery in the ship was not broken, for an excellent luncheon, with all necessary plates and glasses, was soon ready for us.

For the rest of the day it blew and rained, and next morning was blowing still, with no sign of a change for the better in the weather, and, as time was limited, we could only bow to superior force, accept a defeat, and drive the fifteen miles to Stornoway, passing halfway across a lonely lake with a little island, on which a pair of Great Black-backed Gulls had made their solitary nest.

A rumour, soon after, found its way southwards, through the *Oban Times*, brought by the passengers on board the *Hebridean*, which effected a landing, not many days after our unsuccessful attempt, that since the St. Kildians had then last held communication with the outer world, two strange birds, "like Razor Bills, but twice the size"—a fair rough description of the Great Auk—had been seen by more than one of the islanders. Stories to much the same purport have, during the last eight or ten years, come from the coast of Norway.

There is nothing impossible in the idea that the bird may be still in existence somewhere, and nothing more probable that, in such case, one may sooner or later find its way to St. Kilda, where it was once common.

Martin, writing in 1697, mentions it first in the list of sea-fowl visiting the island—"the stateliest as well as the largest of the sea-fowl here. He comes," he writes, "without regard to any wind, appears the first of May and goes away about the middle of June." Sixty years later its visits were becoming less regular, but it was still a familiar bird. Macaulay, who visited St. Kilda in 1758, as missionary from the Society for Propagating Christian Knowledge, which, as he tells in his preface, took "a peculiar concern in the people of that Island," says, that he had not himself, during his stay, "an opportunity of knowing it." The St. Kildians do not, he adds, "receive an annual visit from this strange bird as from all the rest on the list and many more. It keeps at a distance from them. They know not where for a course of years. From what land or ocean it makes its uncertain voyages to them is a mystery of nature."

There is something dramatically appropriate in what, until rumour gives way to something more definite, must be considered the last appearance of the Great Auk. The story is told by the authors of the "Fauna of the Outer Hebrides," on the authority of Mr. Henry Evans, of Jura, who learnt it from old men who had known the chief actors in the tragedy.

In the month of June, in or about 1840, three men

from St. Kilda landed on a neighbouring rock, the Stack-au-Armine. Half way up the Stack they found a strange bird.

" Prophet-like that lone one stood "

—the last of his race. One of the three men caught it by the neck, while the others tied its legs. For three days it was kept alive, but on the fourth day a storm sprang up, and it was sentenced to die as a witch. Solitary, and misunderstood, the bird fought hard to save a species from extinction, biting nearly through the ropes that tied it, and was not killed until it had been "beaten for an hour with two large stones." An Auk's egg was sold by auction in Stevens's Rooms, in 1888, for £225. It would be curious to see what Lauchlan Mackinnon, who, if living, still would be little over 80, could now get for such a captive, well advertised for sale, alive.

Most of the early writers mention, as a characteristic of the "Gair Fowl," a "hatching spot—a bare spot," that is, writes Martin, "from which the feathers have fallen off with the heat in hatching." The peculiarity is noticed in its Gaelic name, "An Gerrabhal," translated by Mr. A. Carmichael, "the strong, stout bird, with the spot."

The presence of "the spot" on the breast of the Great Auk is the more worth noting, as a story told by earlier travellers of a nearly allied bird, one of the Great Penguins of the Southern Hemisphere, but not, perhaps, very generally believed, has lately been found to be true.

The bird has between its legs a fold of bare skin and muscle, hidden under the breast feathers, forming what is practically a perfect pouch, in which it can, and, as proved by the officers of the *Challenger* Expedition, actually does, carry its egg.

The town of Stornoway, when we reached it, was astir with men and girls busy with preparations for a start southwards by a special steamer to sail in the small hours next morning.

The fishing, just over, had been another instance of the "glorious uncertainty," which adds half its zest to idle sport, but is a terrible fact to be reckoned with by those whose provision for wives and families is dependent on their earnings.

The Barra herrings, which, for some reason, which no one seems quite able to explain, fetch higher prices in the foreign markets—Russian especially— than any others, had failed entirely. The fish landed at Castle Bay were little more than a third of the average catch of the last eight years and less than a quarter of the catch of 1885. In Stornoway, only a few miles to the northwards, the take was almost unprecedented.

A thousand girls, imported by the curers from Yarmouth, Grimsby, and other southern ports, engaged for the season at 16s. a week—fish, or no fish— with a free passage from and to their homes, were busily employed for weeks cleaning the fish, and, though the work was over before our return, the air within a circle of half a mile of the town was still heavy with the faint oily smell of herrings.

Our steamer, which carried also its full contingent of fishermen and girls, sailed at midnight. By noon the following day, after a bath and breakfast at Strome, we were rocking along the Highland line pitching and tossing in a manner which would scarcely have imperilled the reputation of the *Jackal* as the liveliest sea-boat in Her Majesty's Navy.

The hills and woods in the soft monotonous green of early summer, looked smaller, but scarcely less beautiful, than in the reds and golds of autumn, more familiar to southerners. Black-headed Gulls flew peacefully over the Lochs, and every now and then a Heron lifted a long stiff neck from a reed bed without troubling himself to rise. Once as we rounded a corner we came suddenly upon two fine stags within a hundred yards of the wire fence which shut off the line. They lifted their heads for a moment in perfect unconcern, and before we were out of sight were browsing again; the pale coloured patch near the tail which, more or less distinctly marked, is a characteristic of most of the deer tribe, showing conspicuously as they stood with backs towards us in the sunshine.

In most of the more genuinely wild species, the patch round the tail is more clearly marked than is the red deer; and, if the conclusions of Mr. Wallace are correct, plays an important part in the preservation, at least of the more gregarious families, serving as a mark of identification by others of the same species, and as a signal of alarm when one of the herd scents danger.

The scud of the rabbit serves, we are told by naturalists, in part at least, the same purpose. It is inconspicuous so long as its owner is quietly feeding in the dusk, but the moment he is frightened and starts post-haste for his hole, it is waved as a white danger flag for the benefit of the many who are pretty sure to be feeding near.

Altruism—clumsy and un-English as the new-fangled word sounds, it fills a gap in the language—is, at least throughout the lower orders of creation, the law of Nature. The individual must suffer for the benefit of the species. The white tail of the rabbit is an alegory. Its owner affords a clear shot in the dusk that his brothers may be warned of danger before it is too late.

Of deer, and of rabbits, it is true, as of men, that "None liveth to himself, and none dieth to himself."

It was tantalising after three years' planning to have been brought at last within fifteen miles of the Holy Land of one's longings and then to be turned back. But, happily for most of us, in this world of light and shade, the minor trials of life, at least, commonly bring with them the element of compensation which—

"Gives even affliction a grace,
And reconciles man to his lot."

To return to London without having caught a glimpse of the Peak of Conagher was a disappointment. But, perhaps, on the whole, it was well.

As one gets on in life the pleasant illusions of youth which survive become fewer, and it is something to have spared even one of them.

If the wind had slept for another day, or if it had awakened in any other quarter, St. Kilda would probably have been now for us a hilly island, inconveniently situated; disfigured by unromantic cottages with corrugated iron roofs; with a population a little spoilt by the visits of excursion steamers; with birds on the cliffs more plentiful even than we had pictured them, but with very little else to distinguish it from many another island more easily to be got at.

But the wind roused itself to blow freshly from the south-east, the one quarter to which the bay and landing-place are hopelessly exposed, and St. Kilda is still the Garden of the Hesperides of boyish dreams—the inaccessible, enchanted island where the reckless fowler let go his rope as he gathered eggs in the recess into which he had swung himself under the over-hanging rock halfway down the cliff, and nerved himself to spring six feet over eternity and catch it as it swung loose in the wind; where the father, when he saw the rope, on which he hung below his boy, chafing against the sharp edge of the rock above, and no longer strong enough to carry both, drew his knife, and cut his own weight off to save his boy; where the miracle of the feeding of the Israelites in the desert, and Elijah by the Brook Cherith, is still every year renewed, and the meat of the people brought on the wings of birds, at the ordering of the God of Nature, who teaches "the Stork in the Heaven to know her appointed season," and Fulmars and Gannets, as well as "the Turtle, the Crane, and the Swallow, to observe the time of their coming."

In Dutch Water Meadows

"Where the pent ocean, rising o'er the pile,
Sees an amphibious world beneath him smile.
The slow canal, the yellow blossomed vale,
The willow tufted bank, the gliding sail."
—*Goldsmith.*

IN these materialistic days it is at the bidding of the poet only that the shadow of the sun-dial moves backwards. If the more glaring the improbabilities in the face of which the miracle is performed, the greater the genius of the worker, among the greatest of the poets and poems of recent days must be Goldsmith and his "Deserted Village." Sweet Auburn, with its garden-flowers growing wild, and Bitterns returning to nest in spots where once villagers had danced and talked local politics, is as real to most of us as Charing Cross, though we know well enough that as "wealth accumulates," trim gardens, instead of running to waste, push out in every direction. It is the Bittern which is giving place to man, and not man to the Bittern; and if we want to see anything of these and other waders which only a generation ago were common in England, we must turn our backs on home, and look to countries where unreclaimed land is, in proportion to population, greater than it is with us.

Slowly or quickly, the same process of extermination is going on everywhere. The Dodo and Great Auk have disappeared. The Ground Parrot, the Kiwi, and the Bison are disappearing. The northern half of Texel, not long since the chief of European breeding stations for long-legged birds, is drained and ploughed, and is Eerland—" Eggland" no longer in anything but name.

But places are still, under good guidance, to be found where the shadow seems to have stood still, and where—as in Prospero's Isle—the air in spring and early summer is "full of noises, sounds and sweet airs," as if all the electric bells and flutes in world had taken flight together, and where the intruding listener's ears are all but boxed with the wings of indignant Peewits and Redshanks. It was in such a spot that we found ourselves on the 3rd of June.

We had crossed by Rotterdam and spent an afternoon in the Museum at Leyden, inspecting, under the guidance of Dr. Jentinck, the Director, some of the most precious of the treasures there. A Duck and other birds believed to be unique, or almost unique, examples of extinct species; the Pigmy Hippopotamus from St. Paul's River; the Banded Bush Buck, unknown until Herr Bütekoffer lately brought it from Liberia, excepting from two imperfect skins—one of them made up into a native African hunting-bag—from which had been evolved and fairly accurately figured an undiscovered antelope*; the rare Flat-nosed, Two-horned Rhinoceros; a Great Auk in good preservation; a huge and almost perfect Epiornis egg, bought from a Frenchman for a thousand guilders—in our money about £80—something less than a third of the price paid not long ago for an Auk's egg sold by auction at Stevens' rooms.

After a five o'clock *table d'hôte*, with a *menu* to remind us that we had crossed the Channel, a *water souchet* of Perch with resplendent fins, served with boiled parsley, Chicken with *compôte de fruits*, &c., we had made the most of the remaining hours of daylight by driving out beside canals and ditches glorified by a golden sunset, and through copses ringing with the songs of Ictarine Warblers and Nightingales, to see a Stork's nest, the pride of a neighbouring village. It was on a cartwheel on a high pole in a meadow, near the church, carefully fenced in. Both birds were at home. As we came up, the female, who was "sitting," lifted her head for a minute, and, coming to the conclusion that we were harmless, settled down again. Her mate rose and sailed slowly round the meadow, to return again very soon, and when we drove off stood on one leg, a feathered St. Simeon Stylites on his column, sharp cut in purple shadow above the trees, beside the low-spired tower, against the evening sky.

Storks are becoming much less common in Holland

---

* A figure of the Banded Bush Buck, with horns and hoofs judiciously hidden by foliage, as neither of the skins had heads or legs, was published in 1841 in the "Zoologia Typica," by Louis Fraser, naturalist to the Niger Expedition than they were a few years ago, and though occasionally we saw a stray bird or pair, this, and one other of which a passing glimpse was caught from the train, were the only nests we saw.

We had steamed next day in a spanking breeze from Helder, the Portsmouth of Holland, across the Dutch Solent, through a fleet of Texel trawlers, which lifted at one moment their heavy bows clean out of the water, and the next moment dipped until half hidden in clouds of spray. We had spent a quiet night in the cleanest and sleepiest of little inns, and—after an early breakfast in a room looking out on a miniature square paved with bricks on edge, in deep shade, excepting where dotted with the few specks of almost tropical sunshine which found their way through the foliage of twenty-nine closely planted lime trees in full leaf, resonant with the notes of warblers and starlings—had been driven with a pair of fresh horses for some miles along the top of a wall like the back of a knife, on the one hand the sea, on the other, apparently at lower level, ditches and meadows. From the top of the wall we had dropped down suddenly to an inland country, to be reminded that the sea was not twenty yards off, as every now and then the sails of a fishing-boat showed over the green banks which we skirted.

For another mile or two we had jolted along a cart-track, till, our coachman having lost his way, we were brought to a full stop by a ditch and rail. At last we had succeeded in finding and introducing ourselves to the agent, who, with the kindness almost invariably shown by the Dutch to strangers, had given us leave to wander at will over the land under his charge.

It was a "polder," a wide tract comparatively lately

reclaimed, intersected in every direction by ditches at right angles; in parts dry and cultivated, in others, on the seaside especially, still in a half swampy state.

It was here, where the deep green of the grass was in places broken with sandy strips and muddy inlets, and in others bright with thrift and white and yellow blossoms of different kinds, that the birds and nests of which we were in search were most plentiful. The air was filled and the marsh and meadows alive with noisy Redshanks and fairy-like Terns, the "Common" and the "Lesser." Oyster-catchers, affected with the usual low spirits of their race, lolled about in disconsolate attitudes, or rose with a melancholy piping as we came too near them; and, where the grass gave place to pale-coloured mud, Kentish plovers elsewhere rare, looking more like little balls of living sand than birds, trundled themselves at a great pace out of our way along the water's margin.

For these and many others, any of which would elsewhere have been worth a special pilgrimage to see, we had no eyes to spare.

We were in one of the chief of the few remaining summer-homes in Western Europe of the Avocet, once common, now practically extinct, in England.

One of the last of our old-established colonies was at Salthouse, on the Norfolk coast, and was, according to tradition, destroyed in the first half of this century for the sake of the birds' feathers, which were in request at the time for making artificial flies.

No one who has only seen an Avocet stuffed can form any idea of the grace of outline and motion of the living bird; nor of the bewildering permutations and combinations of its zebra stripes of black and white.

For half a moment, as it settles, the bird is still, and you see two distinct horseshoes of jet on a snowball. Before the roughest sketch is possible the position of the restless wings shifts and the horseshoes meet and open into a double heart, one inside the other. It rises, breast towards you, and you see a bird, pure white excepting at the wing-tips, which look as if dipped in ink. It turns sharply off, with the everlasting "Kiew! kiew!" and you seem to be looking not at a bird, but an overgrown "Bath-white" Butterfly.

At last you have had one quiet before you long enough to be satisfied at least that the tail is black, and are hurriedly scratching a sketch accordingly, when the black flies up on the tips of the wing and the bird is off, turning towards you a tail of the purest white.

They were very plentiful, and wonderfully tame. We must have seen something like fifty on the one corner of the polder, to which they seemed mainly to confine themselves, and where we found both eggs and young birds.

As we lay for luncheon on our macintoshes spread on a patch of thrift, not far from the water's edge, the old birds played and fed close by us, swinging sideways, their slender turned-up beaks—like strips of bent whalebone—splashing visibly at times with the strokes, and ran bent forward through the water, sometimes breast-high, with a quick, jerky, and rather laboured step, the position of the body and action suggestive of a long-legged, paddling child in a great hurry to get a shrimping-net on shore.

The neck, as the bird ran or fed, was commonly

drawn backwards with a curve like the droop of the dewlap of a cow. The young birds, of which we caught two in different stages of growth, mimic their mother's steps as they run, and could be identified by this even without the fascinating little baby *nez retroussé* which makes mistake impossible. One, a little striped puff-ball, which could not have been many hours out of the egg when we found it, feigned desperate illness rather too well, and was all but pocketed as past all hope of recovery. But when left alone, unobserved as it supposed, on the grass for a few minutes, rose quietly, and after creeping slowly through the stalks for a foot or two, reached a sandy "grip," when it set off running at a pace miraculous for so small a creature.

The legs of the old birds are bare for some inches above the joint, which is very prominent, and are of a silvery grey, not many shades removed from Cambridge blue, and are more slender than in the pretty picture in Lord Lilford's book.

In flight the legs are tucked tightly under the tail, of which, when the bird is in the air, they seem a part. The body is exceptionally flat, so much so that an Avocet flying looks as if it could have no stomach.

In spite of their slender make they are courageous, and if offended fly at more stoutly-built birds. A couple of days later, on another marsh, we watched for ten minutes or more one of them vociferously attacking a Black-headed Gull, who—perhaps because it had been sucking eggs, and conscience had made a coward of it—was evidently very anxious to shake off its pursuer. The Avocet circled upwards like a Falcon, and swooped with a scream again and again at the Gull from above, never, so far as we could make out, actually striking it, as the scarcely heavier Richardson's Skua would have done if offended, but swerving sharply to the right or left when within a foot or so of its enemy.

Not far from the flowery slope on which, "at ease, reclined in rustic state," we sat to lunch and meditate, was a ditch rather wider than some—one of the arteries of the polder. The mud of successive cleanings had been thrown out on the side nearest to us, and had dried into a bank a little above the general level. It was what in old days was known in the Fens and Broads as a "hill"—a gathering-place of Ruffs, birds which once, like Avocets, were common in England, and are now scarcely less rare.

More than once we counted nineteen or twenty of these curious birds together on the hill, and many others constantly came and went. Much has been written of the fights of Ruffs, which—unlike most, if not all, the rest of their class—do not pair, but are, like Pheasants and Barndoor Fowls, polygamous.

But, perhaps because questions of precedence had already been settled, or perhaps because it was not until towards the afternoon of a hot day that we found them in any numbers, we saw nothing ourselves to justify their distinctive epithet, "Pugnax."

Every now and then one of the party rose, bowed, and pointed his beak at a neighbour, who acknowledged the compliment in the same manner. The two, to borrow a phrase from *Punch*, "flashed their linen," ruffling their frills to make them show to the greatest advantage, bowed a second time, and settled quietly down again. There was occasionally a little momentary excitement, as another of the privileged circle dropped in, looking as he flew with ruff closed like a little pouter pigeon, but nothing like quarreling. Everything was done with quiet decorum, and the general effect was more that of a select club window in St. James's Street on a June afternoon than of a duelling ground.

No European bird, probably, varies in colour to anything like the same extent as the Ruff. Of the many we saw no two were nearly alike in plumage. One that we watched from close by with a glass was noted as having a chestnut ruff with a black face. Another had an almost pure white ruff and chestnut back. A third had a white ruff, broadly tipped with black, and a back of the sandy dun of a little ringed plover. A fourth had a ruff of black and white in diamonds, like a shepherd's plaid. Two were, or appeared to us to be, ruffs and all, whole coloured, the one a neat uniform slate grey, the other cinnamon. Another, a great beauty, had a ruff of the darkest glittering purple shot with blue. The eggs of the Reeve are smaller and more highly polished than those of the Redshank, which they generally resemble, and are commonly more richly and uniformly spotted. The age at which the Ruff in a wild state justifies his name and dons his Elizabethan collar, is a little doubtful; but there is not much doubt that it is not until he is at least two years old.

Our attention had been so much occupied with the larger and more obtrusive birds, that we had not much time left for the little birds. But among many which elsewhere would have been remarkable were a pair of Blue-headed Wagtails, with breasts of vivid yellow, and a third Wagtail almost pure white. The last was in company with a female of the ordinary "pied" species, of which it was probably an accidental albino variety. We saw it twice at an interval of an hour or two, at the same spot, beside a ditch where it probably had a nest.

To the south of our polder lay a narrow tract of sand-hills which, seen through the shimmering heat from the dead level of the old sea-bottom, looked like a distant mountain range. It was a pleasant change, after having been scolded for hours in shrill tones in every key, to climb the first ridge and drop into another world. Excepting when, every now and then birds, singly or in pairs, passed overhead, the noisy tribes of the flat lands and ditches were left behind, and not a sound was to be heard louder than the gentle rattle of the Dry Bent as it moved in the breeze, the trill of one or other of the little warblers, which in summer time are commoner, perhaps, in Holland than anywhere, or the song of a distant Lark in a sky, the faint blue of which blended perfectly with the pale browns and yellows of sand and bleached grasses.

As we sat among the sand-hills enjoying the calm, three Hares, smaller and darker than our own Lowland Hares, followed a few minutes later by a fourth, passed within ten yards without noticing us.

Our first day in Texel was past and gone, a pleasant recollection only. A second and a third, as pleasant, followed, to fly as fast.

In a slushy water-meadow, eight or nine miles from our first hunting ground, we stood in the middle of colonies of the Black and "Common" Terns, which bred in sociable company with Godwits and Black-headed Gulls.

We had wondered at the courage of the slender Avocets when man or bird approached their nests.

They were cowards compared with the little Black Terns, which, as we stooped beside their eggs, dashed at us with the recklessness of Skuas.

They are beautiful birds as seen from below, with slate-grey wings and bodies of shining black, shorter and smaller, but proportionally stouter than their fork-tailed cousins, the Common Tern.

The nests of the Godwits, of which we found more than one, unlike those of the Avocet, which lays its eggs in a bare hollow of trampled turf, were thickly lined with dry grasses.

The birds themselves—" which, by-the-by, were once," writes Sir Thomas Browne,* "accounted the daintiest dish in England, and, I think, for the bigness, of the biggest price"—with their long beaks, were conspicuous and unmistakable at almost any distance, in their bright summer dress of brownish red and white. The female, as with the Hawk, is the larger bird.

In the deep blue water of an irregular natural pool, in striking contrast to the formal artificial ditches of the drained lands, we counted at one time ten separate species of water birds together, and not unfrequently had five or six kinds in the field of the glass at one time. Nearer home we crawled through copses hedged with tall green reeds, to watch the Ictarine Warblers, seldom seen in England, but here common. The capricious Nightingale plentiful almost everywhere on the mainland opposite, is, we were assured, almost if not quite unknown in Texel.

It is always interesting to trace in every-day life survivals of old ideas and customs underlying modern thoughts and habits. It is not often that

* On "Norfolk Birds," vol. iii. of Works.—*Bell.*

the old and the new are to be found in such grotesque conjunction as in the head-dress of the well-to-do Dutch farmer's wife of to-day. But when family jewels and old lace come into collision with fashion, Greek meets Greek, and neither gives way in a hurry.

The picturesque polished silver head-plates under the pretty cap of fine lace or blue silk gauze, and gold face ornaments which may have formed part of the "Ladies' Subscription Fund" towards the cost of flooding the country for the relief of besieged Leyden, or have been buried for safe keeping in the days of "the Spanish Fury," are still to be commonly seen in Sunday wear, but scarcely ever now without a vulgar parody of a Paris bonnet of a year ago like a mocking imp straddling on the top.

The blue gauze cap is worn only by Roman Catholics. The same distinction of creeds is marked also by the colour of the awnings of the family carriages, which, with their high carved tail-boards, look like Old World ships placed, stern foremost, on wheels. It is a fairly safe assumption, though less universally true than was once the case, that the farmer's wife and daughters who look out at one as they drive by from beneath a white hood, are Catholics—from beneath a black hood, Protestants.

But time is short. Almost before we can realise what it is that we have been looking at, another slide is in the lantern. The bright greens and pinks and blues and yellows of the Dutch polders, and the softer tints of the sand dunes behind, fade on the sheet, to re-arrange themselves in more sombre tones. The windmills and heavy pyramids of straw thatch on stunted walls—farmhouses and barns in one—" dissolve," and give place to shops and clubs.

The changing scene has shifted once again to London.

London Insects.

"Come take up your hats, and away let us haste
To the Butterfly's ball and the Grasshopper's feast;
The Trumpeter Gadfly has summoned the crew,
And the revels are now only waiting for you."
—*Roscoe.*

IN Alphonse Karr's little book already quoted there are two chapters with very suggestive, if not very elegant, headings, " Sur le dos" and "Sur le ventre." Lying on the grass, and looking first down in the strong light into thickets of thyme, and moss forests of miniature palms, and tree ferns, to watch the important doings of insects which live there, and then turning over, and looking with shaded eyes through the gaps in the branches of the larger trees and floating gossamer threads at the clouds and clear depths of sky beyond.

It may depend on the turn of one's mind whether,

used independently, the telescope or microscope sets one thinking most, but together they must make the least imaginative of us feel what a very little space it is after all that man, with all his many inventions, occupies in creation.

> "Vast chain of being! which from God began.
> Natures ethereal, human—angel, man,
> Beast, bird, fish, insect. What no eye can see,
> No glass can reach. From infinite to thee,
> From thee to nothing."

"Glasses" have much improved since Pope wrote these lines 150 years or so ago, but have brought us as yet no nearer an upward or downward vanishing point of possible life.

Each increase of power of the telescope has, on the contrary, only revealed fresh worlds beyond, and in the microscope only brought us nearer something like ocular demonstration, that there is truth in the homely lines which tell that

> "Little fleas have lesser fleas upon their backs to bite 'em,
> And those fleas have other fleas, and so ad infinitum."

In which ever direction we look it is the same. More knowledge of nature means more consciousness of life all round us.

If the exploring voyage of the "Challenger" has proved any one thing more clearly than another, it is that the old idea that ocean life ceased at a certain depth has no foundation in fact.

There is one link in the chain of life, and only one, which Londoners have ample means of studying in the open air—the order of "insects." Beasts to live naturally require solitude and room to wander and breed. Birds scarcely less. But for many of the most interesting of the insects a single leaf is an estate, and a park or garden in a square means space unbounded; and so, as might be expected, they swarm in all directions in London as thickly as anywhere else.

Perhaps, indeed, there is no place in the world in which with good eyes and a little patience such a curious collection of miscellaneous living insects from every part of the world might be made as in the neighbourhood of the London Docks. They come, whether we will or not, as stowaways from every port with which our ships trade—from every part, that is, of the habitable world.

It was only a very few years ago—as late as 1877—that the help of the three Estates of the Realm was called in to prevent an invasion of a small yellow and black-striped beetle, scarcely bigger than a Ladybird. An Act of Parliament was passed, two Orders at least, with pains and penalties, were issued in the Queen's name, and something over 10,000 broad sheets printed and circulated broadcast throughout the country, giving coloured pictures of the miscreant, life-sized and magnified, in every stage of its existence, and announcing in colonial English and large print that "the country around the town of Ontario, Canada," was "swarming with the Colorado Beetle," and that the Canadian Minister had reported that not only did it "move by flying and by navigating, so to speak, smooth water, but also travelled on common vehicles, railway carriages and platforms"—most alarming of all—"and on decks of vessels, &c., especially during the months of August and September."

Solomon on his flying carpet, with his royal escort of hoopoes, would be a scarcely more startling party to meet than a company of Potato Bugs crossing the Atlantic on a railway platform.

There is a tradition that an insect even more hateful, the common bug, was unknown in England before the great fire, and that it was first imported with the fir wood brought over wholesale to rebuild the city.

It is impossible now to say how far the story may or may not be founded on fact; but as most of the timber used in building before 1660 was most likely native-grown, there is nothing improbable in it. The passage in Matthew's Bible, published a hundred years before the fire, which gives as the promise of the 91st Psalm that we shall "not be afraid of the bugs by night," cannot be quoted as any proof that they were known then, as it is only in later days that the application of the word has been limited to the particular "terror by night" which now monopolises it. But if there is any truth whatever in the story it ought to make all of us who are fond of a potato, Irishmen in particular, very grateful to the authorities whose energetic action—not taken until Germany had already been successfully invaded—has so far succeeded in keeping the Colorado Beetle from landing in any force, if at all, on the coasts of England or Ireland.

"Insects,"—creatures made in sections,—insectæ ("entomology," the science of the study of insects, is a compound of the same word in a Greek dress), stand about a quarter of the way up the great ladder of intelligent life, which has man on the highest rung which we can see clearly now, and its foot on the uncertain ground—"that low department of the organic world from which the two great branches rise and diverge," where creatures without heads or nerves or apparent consciousness, but holding by patent conferred by learned men, who have spent their lives in studying them, brevet rank as animals, mix in undistinguishable confusion with so-called vegetables which move and act as thinking beings.

They are the lowest class of the third great natural order, the *"articulata,"* the jointed animals which have their supports answering to a skeleton outside, not like boned animals, inside the soft fleshy parts.

Below the insects stands the whole order of the "Radiata"—star fishes, &c., which, made in rays running out from a centre like the spokes of a wheel from the axle, give the order its name—coral makers, animal flowers and jellies, and tiny revolving wheels and balls.

Below, or side by side with the insects, according to the point of view from which we may prefer to look at them, stand Crabs and Lobsters and Worms, and above them, Spiders.

Above these stands, with its many subdivisions, the order of the Molluscs. Soft-bodied boneless creatures, with a perfect system of circulation of blood, such as it is, colourless stuff. Never "articulated" or jointed like the class below, with endless varieties, from the Slugs and Snails which give London gardeners opportunities enough of studying the family in its humblest members up to the Cuttlefish and the Nautilus—

"The sea-born sailor of a shell canoe."

Higher still comes the noblest order of the Backbones running through all its grades—Fish, Reptiles, Birds and Beasts, till actual knowledge is brought to a full stop for the present at Man.

With the help of Sir Brydges Henniker and the editor of the London Postal Directory it might be possible to make out a list of the human families inhabiting London and their occupations. But to do the same for its insects would be quite impossible. The outside which any prudent man would attempt within the limits of a chapter, is to mention one here and there—enough to show that, as with the Birds, we have typical representatives in London of all the principal natural groups.

But even this is less simple than it seems, for at starting we find ourselves face to face with questions which have vexed naturalists from the days of Aristotle, 2,000 years or so ago, and probably from much earlier times, for Solomon wrote of creeping things, and Moses certainly had made some study of them. What are to be called "true insects?" and is there any one way in which they fall more naturally than another into groups?

The decision at which the learned have now arrived, with something like one mind, as to the answer to the first question—What shall an insect be?—is that *true insects* shall not, as was once held, be all such articulated (outside-cased and jointed) animals as may be cut into sections—an arrangement which lumped up Crabs and Lobsters, with Gnats and Butterflies—but *such articulated creatures only, with or without wings, as may, at one time at least of their existence, consist of exactly thirteen segments, and may in the perfect state find themselves masters of six legs, no more and no less.*

When one is told that this short definition covers almost everything in England—in the known world, indeed—which we should be likely in common talk to speak of as "*an insect*," with the exception of Spiders and Centipedes, and as Swift describes it—

"That curious creature men call a woodlouse,

("Slate beast" is its name in the Highlands)

"Which rolls itself up in itself for a house,"

one scarcely knows whether one ought to feel most astonished at the wonderful likeness in unlikeness running everywhere through Nature, which makes such generalisations possible, or at the labour which a conclusion of the kind represents.

Why should every Butterfly, Beetle, Moth, Fly, or Flea wherever found—north, south, east, or west—be built up of just thirteen segments, and have just six legs, and how many centuries of quiet work of patient, observant men—living and dying many of them absorbed in the one favourite study—has it taken to find out with something like certainty that such, improbable as it sounds, is the case?

To the second question all sorts of answers have been given at different times, each seeming satisfactory at the moment, but most of them to be written only on sand and washed away by the rising tide of fuller knowledge.

There are at least three ways in which the six-legged insects may be grouped. One natural division is according to their manner of feeding—some suck, others chew or munch.

The first are spoken of in scientific books as "*Haustellate*," the last as "*Mandibulate*," but as this arrangement only gives two classes, it is not of much use. Another way of dividing them is according to the changes which they pass through before reaching the perfect state.

Moths and Butterflies, and the many other insects which, before they can fly, are at one time Caterpillars, or Grubs of some kind, and at another Chrysalises— each stage to all appearance quite unlike the others— are classed as undergoing "complete metamorphosis," and are called "Metabola."

Others, such as Earwigs, Green Flies, and Cockroaches, which pass through less startling changes, are classed as undergoing either "incomplete" or "no metamorphosis"—"Hemimetabola" or "Ametabola."

The difference between the three classes thus divided seems very wide; but is, in reality, less than it appears at first sight. It lies mainly in the difference in the stage of growth at which the insect is born or hatched.

The "unchanging," or only "partially changing" insects, leave the egg in more or less advanced stages of development, and reach the perfect state rather by gradual growth than by any sudden alterations of form; while those described as undergoing "complete metamorphosis" are first hatched in such an imperfect form that, after eating and growing for a time, they are practically sent back again to the egg. In this second torpid egg-stage the soft fleshy parts are hardened, and wings and other high organs developed, of which there were no traces when the insect left the first egg. When this is done the chrysalis skin is cracked and thrown aside as useless, like the eggshell of a Chicken. The difference between insects undergoing "no metamorphosis" and those undergoing "partial metamorphosis" is merely in the stage of advancement in which the egg is left. Both leave it in a more developed condition than the classes of insects undergoing "complete metamorphosis."

But as this arrangement, though a little further reaching than the first, gives only three classes, and these with no very clear lines between them, it is not of much more value. For all practical purposes the only possible classification of insects yet worked out is according to the nature of their wings. "Those Hexapod insects," writes Professor Owen — the greatest living authority perhaps on all such matters —"which are devoid of wings, are called *Aptera ;* those with two wings only are the *Diptera*. All the rest have four wings. The *Lepidoptera* have four scaly wings ; the *Hymenoptera* have four veined wings, crossing each other when at rest ; the *Hemiptera* have one pair of wings partially thickened, and called hemelytra ; the *Orthorptera* have one pair of wings wholly thickened, the other folded lengthwise ; the *Coleoptera* have one pair wholly and much thickened, called elytra, and the other pair folded crosswise; the *Neuroptera* have four reticulated wings; the *Strepsiptera* have one pair of wings rudimental and curled up. In the *Aphaniptera*"—which, by-the-bye, are not an order by themselves, but only a class of the so-called wingless order—" both pairs are rudimental and functionless as wings. Of these orders the first five are 'haustellate' ; the next four are 'mandibulate.' The Aptera are 'ametabolian' ; the Hemiptera and Orthoptera are 'hemimetabolian'; the remaining orders are 'metabolian.'"

"These characters," Professor Owen adds, "briefly and succinctly express the highest generalisations, as yet reached, relative to the Hexapod Insecta."

But here, as in every attempt of the kind, the more perfect the work the more one is made to feel that classification, absolutely necessary as it is as a step towards progressive knowledge, is at best a purely human invention, and that there is no such thing as a hard-and-fast line anywhere in Nature, where all is gradual. We can find out without much difficulty —when the gas lamps are lighted—they are human institutions, but it would puzzle the wisest of us to say exactly when it is that the "crimson streak" on the Serpentine "grows into the great sun." No wings, two wings, and four wings are, we can see, the general characteristics of large classes, but at what precise spot the separating lines are to be drawn on paper must be, to a great extent, a matter of fancy. They may be, in most cases, moved up or down without its mattering very much. As already mentioned, among the wingless insects is a class—the "Aphaniptera"— which has "scales representing rudimentary wings." Most of the "two wings" have a rudimentary second pair, known as "balancers," behind their more perfect first wings, and a whole order taking rank among the four wings—the "Strepsiptera"—have nothing better to show, as front wings, than the miserable little "screwed up" apologies which give the order its name.

Unfortunately, however much we may have cause to lament the scarcity of the more showy Butterflies in London, we cannot complain of any want of specimens of the first great order of insects, the "wingless." There are three classes of the order, all most objectionable, and the less said about any of them, perhaps, the better. The class, already referred to, which, having little scales to mark where wings should be connects the order with the flying insects, is certainly not the worst.

Its most familiar species is the Flea. Bad as it is, with its "double lancet mouth," in one respect the Flea ranks high in the moral scale.

The infant members of colonies of Bees and Ants are, as we all know, fed by their elders. But it is never the actual mother who does the work. With few if any known exceptions, Fleas and Earwigs are the only common English insects which feed their own young themselves, and of the two the smaller is perhaps the more deserving of praise, for she, on her foraging raids, risks her life in a way which an Earwig never needs to do. There are only two allusions in the Bible to Fleas. Both are in David's appeals to Saul, and are suggestive, not of any evil-doing, but only of the miserable hunted life, without a moment's peace and quiet, which the poor things lead. But even with this much to be said in its favour the Flea is an unsavoury subject to write about.

Of the other prominent division of the wingless insects, one individual has immortalised himself—the "ugly, creepin' wonner" which Burns once saw in church, "struntin ower gauze and lace," to

"The vera tapmost, towrin' height
O' Miss's bonnet."

But with this almost solitary exception, the family,

"Detested, shunn'd by saunt and sinner,"

is unmentionable, and, in spite even of Mosquitos and Gnats, it is a relief to pass on to the next order—the "two wings."

Any Londoner who likes may find ample material for a history of one typical family of the two wings — the "Daddy Longlegs" — without any greater physical exertion than an occasional stroll into any of the parks in the late summer. Once there, he will not have occasion to move his chair many yards. They are, especially in warm sunshiny days, when there has been a little rain to soften the upper soil, to be seen in thousands, ladies and gentlemen, and are in many ways satisfactory insects.

To begin with, they are of a reasonable size, sufficiently large to show, without turning one's eyes inside out without any very powerful lens, the curious rudimentary wings, like two sticks with a knob at the end of each, already referred to as a characteristic of the whole order of "two wings." The cutting of the sections is very clear, as is the plated cuirass on the back overlapping the breast-plate, both divided very distinctly into segments which need no glass to count.

They have no smell, which is a consideration, and in spite of the alarming varnished spike which the female carries conspicuously at her tail, they have no sting.

Towards evening there is no difficulty in finding out the use of the terrible-looking instrument.

It is an egg-placer, which can be opened and shut at will, like a heron's beak. The portly-looking mothers-to-be may then be seen by dozens waddling along the grass, or lifting themselves clumsily for a yard or two at a time with flight very different from the maiden dance of the morning. When they come to a suitable place, usually where the grass is thin and a little patch of bare earth is just visible between the blades, they set themselves on end, and either pirouette round and round for a few moments, or make crowbars of themselves, and thump till the spike is far enough into the ground to satisfy them that the cargo of eggs to be slipped through it will be safe. If a little bit of dirt finds its way between the mandibles of the tail-beak of a Daddy—or, to be more correct, Mammy Longlegs—the males end abruptly without any spike—or if anything else happens to prevent them from shutting comfortably, she will reach out one of the hind pair of her six long legs and clear the opening out, and deliberately, in the most comical manner, grasp the points with her foot and pinch and shake them into place again.

The female, when the eggs are laid, is a miserable-looking "shotten herring," back and front plates almost meeting, and probably does not live very long afterwards. Certainly towards dusk one may see hundreds of males under the trees in Kensington Gardens, but has often to look some time before finding a single female.

The larvæ of Daddy Longlegs feed on the roots of grass; they are hatched underground, unlike the Gnats, also "two winged," which are, as everyone knows, hatched on water.

The duty in life of vast numbers of the families of two-winged insects, as, indeed, of most other insects, is to clear away what is most offensive in dead matter, and the way in which they have been fitted for the work is beyond measure marvellous.

Speaking of the Maggots—the larva of the common Blow-Fly—Professor Owen, lecturing in the theatre of the Royal College of Surgeons—neither the man nor the place for random statements—said, "Insignificant indeed do these larvæ seem to be in the scale of nature. Yet Linnæus used no exaggeration when he said that three flesh flies would devour the carcass of a horse as quickly as would a lion. The assimilative power is so great in the meat maggot that it will increase its own weight 200 times in 24 hours."

It is not easy to say which is most astounding, the

self-reproductive powers and quick growth of many of the commonest insects, or their powers of consumption.

It was a locust who told Mahomet that his tribe were the army of Allah, and that had it pleased Him to fix the number of their eggs at a hundred instead of at ninety-nine a-piece, the earth and all that was in it would have been consumed.

It is not necessary to look beyond our London flower-boxes—green and bright with blossoms one day, and a week later, little better than dry skeletons, or with every leaf riddled as if it had been a target for snipe shot practice—to find proof enough that Moths might boast as much.

Two years ago a couple of homely-looking brown Moths, the common Cabbage (*Mamestra brassicæ*), found their way into a conservatory, the ornament of a London house. The owner was weak enough to be flattered with the thought that the Moths had chosen the place as the best imitation of the country they could find in London, and with some sort of foolish idea that their presence added to the rural charms of the fern banks, in a fit of mistaken tenderness (perhaps of vanity) they were left undisturbed, and made themselves at home for a day or two.

It was the husbandman warming the snake in his bosom; Sinbad the sailor giving a lift to the poor old man of the sea; Eve coquetting with the serpent; and the result was in its own degree as disastrous The next year one or two more of the same kind were seen, and the freshness of the country began to leave the conservatory. The following year it was from bad to worse. Human efforts could not stop the mischief, and no little birds were there to come to the rescue; and in spite of a Hecatomb to Dagon—some hundred Caterpillars, all alike, pale underneath, with dark olive-green pencilled backs, thrown into the aquarium for the fishes to fatten upon—pet geraniums were demolished, and some twenty feet of rich rank tradescantia, a plant despised in the country, but very precious to London gardeners for its succulent greenness, which can defy even smoke and dirt, stripped almost to bare stalks. Something over two-thirds of the entire length of a Caterpillar is a disproportioned stomach which the owner must work night and day to fill, nipping away for dear life at whatever green thing comes within reach of his ugly vertical jaws.

The rapid growth of Caterpillars would be incredible if we had not proof. A healthy man takes perhaps 30 years to reach his full growth in height and breadth, and when he has done so weighs probably some twenty times as much as he did when he was born. A Caterpillar will increase its weight proportionately 500 times as much in 30 days. It is difficult to realise what such figures mean; we can get a clearer notion by reversing them. Fancy a baby born of ordinary size growing at such a pace as to weigh when a month old as much as six or seven big elephants together! For the father of a family the idea is too appalling to joke about, but it is no more than would be actually the case if the human animal grew at the rate a wellfed caterpillar will grow in one summer month. Mr. Newport has given from actual observation the weights of the larva of the Privet Hawk Moth—the large, smooth, green Caterpillar, with pink stripes on the side and a horn at the tail—at different ages. On leaving the egg, its weight is not more than about one-eightieth of a grain. When full grown, 32 days later, it weighs from 120 to 140 grains. Take only the lowest weight, 120 grains, it is very nearly 10,000 times as great as the weight on leaving the egg! Apply this to a baby of say 10 lbs.—a good big child, but nothing extraordinary—and we have a very simple sum: 10 lbs × 10,000 = 100,000 lbs., something over 40 tons. And all in one month! The weight of Jumbo when he left England was estimated at something like 7 tons.

But we have strayed from the two-wings to the fairy land of insects, the country of the Lepidoptera—Butterflies and Moths.

A glorious "Red Admiral" was sunning himself outside Buckingham Palace on the 17th September, 1883, shaking the creases out of a very perfect uniform —black, white, and scarlet—evidently just out of the packing case. But we have not often, at least in the central parks or squares, any great number of the brighter-coloured Butterflies, though in London, as elsewhere, the common White Cabbage Butterflies are plentiful. We have become, happily, as a nation, much more tender-hearted than we were in the days of bull-baiting and cock-fighting, and are setting our minds in earnest now against cruelty of every kind. But without necessarily going quite so far as Christopher North, who argued that to give up fox-hunting would be to rob the poor fox of all that made life worth living; the healthy tingle in every limb as he pulls himself together for a start as the hounds are thrown in,— the mad excitement of the first mile's spin across the open, with the pack at his tail—the fun of fooling the huntsman and telling the vixen at home all about it over a good fat hen in the evening:—without going quite so far as this, the man must have forgotten his own boyish delights, who can see without any pleasure half-a-dozen ragged little shouting urchins from the slums of Westminster tumbling over one another in St. James's Park in wild pursuit of a White Butterfly, probably very well able to take care of itself till it meets a White Throat or Cock Sparrow.

It is rather interesting to notice that such Butterflies as there are in London keep to the tops of the trees, more than one often sees them do in quieter country places.

We have a great many Moths of different kinds in London. Judging by the numbers which will hurt one's feelings by flying into the candle as one sits by an open window on a warm night, or finding their way between the globe and chimney of the lamp, where it is impossible to leave them to scorch, as well as by the varieties to be occasionally met with in the parks and streets in the daytime, and by the much too apparent marks of Caterpillars' work everywhere, it ought to be possible to make a large collection.* But, as nearly all are night-fliers, there is not much chance of doing this in the daytime only, and the powers that be are wisely stern in their refusal to create the precedent which the official abhors as Nature does a vacuum, by lending a key of Kensington Gardens or granting leave to outstay the closing hour; and so the privilege of "treacleing" trees in the only really satisfactory hunting fields in town is confined to palace footmen, policemen, and, perhaps, the First Commissioner of Public Works.

* Among the insects caught actually in London by the winner of a prize for Natural History given at a public school were the Leopard, Goat, Ermine, Bufftip, Peppered, "Willow Beauty," and "Brindled Beauty" Moths, and the Small Tortoiseshell, Peacock, and three sorts of white Butterflies. The rare Alder Moth was caught within an easy walk of Hyde Park Corner, on Wimbledon Common.

Less favoured mortals are turned out of the garden with as little hope of appeal as Adam and Eve, and if tempted to cast one longing, lingering look behind, it is only to recognise Michael in Mr. Mitford, and see

<blockquote>
"The gate
With dreadful faces thronged,"
</blockquote>

capped with the blue helmets of the Metropolitan police.

We have, unfortunately, a great many more specimens than we care for of one class of Moth, the little "tineæ," the common Cloth Moth. It is too small almost to be seen except in a good light, but possesses a power, which an electric eel might envy, of galvanising the portliest and most precise of good ladies'-maids or housekeepers into spasmodic jumps and flings, by showing itself near a sable cloak or blanket cupboard.

The typical London Moth is the Vapourer. It is in more ways than one exceptionally interesting. In the first place, unlike most of its kind, which bury their chrysalises and hide themselves until after dusk, the Vapourer is to be seen in all its stages without going out of the way to look for it. The Caterpillar, which is very pretty and curious, has slashes of pink or red, and yellow-pointed tufts of hair sticking up at regular intervals along its back, and longer tufts of darker hair, one perpendicular on its tail, the other two like whiskers, horizontal, one on each side close by the head. It is to be found on the underside of the broad leaves on the branches of the plane trees which hang over the paths in St. James's Park—often with two or three successive out-grown skins, complete with hairs and tufts, on the same leaf with the living Caterpillar. A little later, the same broad leaves are to be seen carrying untidy webs containing a small hairy chrysalis. The perfect Moth—the male, that is—flies in broad daylight, and, as if specially designed for the consolation of country-born Londoners sentenced for any reason to spend an August in London, is to be seen then in numbers plunging about in front of the shop windows in the hottest sunshine, looking—in spite of its beautiful feathery antennæ—less like an insect than a withered yellow beach leaf caught by an eddy of wind.

But the obliging way it shows itself is by no means the only thing which makes the Vapourer specially interesting. It is closely allied to—some naturalists class it with—a family which is of more apparent value to man than all other families of Moths and Butterflies together. It is a silk-spinner, and, in common with many of the family—most notably the Silkworm Moth—the female is practically wingless. It is curious, but not of any great importance to any of us, to know that the female of the Vapourer Moth seldom goes far from the web in which it lay as a chrysalis, and often never even strays outside it. But the same characteristic in the silkworm is of very great importance. For if the female Silkworm Moth had perfect wings, and were free as other ladies of the kind are, to come and go and mate and lay her eggs wherever fancy led her, instead of living and dying content with her own mulberry tree, silk cultivation on any large scale would be impossible. A silk dress would be as rare a treasure as Lady Brassey's feather cloak, and the thousands now employed in the various silk industries would have to look elsewhere for a livelihood.

They are agreeable Moths, too, because they do not waste too long in the chrysalis state—often not more than three weeks; and a boy may bring home a Caterpillar with some hope of really seeing it fly before he has forgotten its existence.

Thus much of the attractions of the Vapourer all of us can see for ourselves by simply keeping our eyes open as we walk under the trees. There is one more point of special interest which most of us must be content to take on trust, but it should be mentioned, if only to show that for any that have eyes and skill for such things, there are ready at hand in the wild nature of London materials for the deepest as well as the lighter studies of Natural History.

Anatomists who have studied the changes of insects (excepting only the resurrection of the dead, of which, as shown by the incriptions in the catacombs, they have been from earliest Christian times the emblem, there is nothing in the whole range of creation so mysterious) have proved that, with very few exceptions, there is a steady growth of all that we should call the higher powers, from the egg, through the Caterpillar and chrysalis, to the perfect insect. The continuous advance, for instance, from the Caterpillar to the higher intelligence of the butterfly, as marked by the fuller development of the nerves in the region of the brain, may be seen very clearly in a beautiful series of ten microscopic sketches, by Mr. George Newport, of the nervous system of the Caterpillar and chrysalis of the Tortoise-shell Butterfly, republished in Professor Duncan's book on the transformations of insects, in which any one interested in the subject may find the results of much deep research in a very attractive form.

The female of the Vapourer Moth, and some

other wingless females, are among the very few exceptions as yet known to this general rule of continuous development. She, when she leaves the pupa state, is said to be actually a less highly-organised creature than she was as a Caterpillar.

The likeness of the chrysalis to a baby girl, "*pupa*," which is the origin of the scientific name, is lost in the long clothes of Modern England; but the sketches given below show sufficiently what is meant. The one is a magnified chrysalis—not, perhaps, that in which the likeness to a child in swaddling clothes is strongest —taken from a sketch reproduced in Professor Duncan's book from one by Réaumer, who was as eminent as a naturalist as in other branches of science, and kept insects in his gardens to observe their habits and changes—the forerunner by a hundred and fifty years or more of Sir John Lubbock and his fellows.

The sketch at the side of the chrysalis is "Nurse Gladstone's Baby," borrowed from *Punch's* cartoon of the 25th August, 1883.

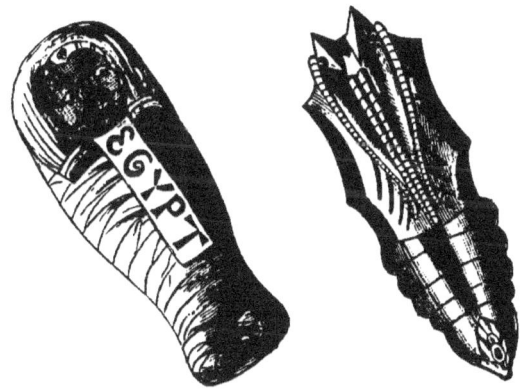

The name for a Caterpillar—"*larva*," a mask—is intended, of course, to suggest the idea of the perfect insect disguised for a time in a form very unlike its own.

If Mr. Worth, or whoever else may be the high priest of the day, should wish for a new combination of grey and brown velvet for a lady's winter dress, with tippets and sleeve-trimmings of fur in two colours, light and dark—the whole relieved, if his copy is to be exact (Nature registers no copyrights in her designs), by a bow of white satin below each shoulder—he will find a model ready dressed, dark gloves reaching to the elbow and all complete, if he looks through a low-power magnifying glass at a Gamma Moth—the "Silver Y" is its other name —which must be as common in Paris as it is in London.

With all due admiration for the present energetic management of the parks and gardens, one cannot help feeling a little low at times on seeing the wholesale clearance—necessary, no doubt, but none the less sad—that is being made of all the shabby old trees.

"Like flies that haunt a wound, or deer, or men,
 Or almost all that is—hurting the hurt."

Caterpillars and small-boring things of all sorts attack, as a rule, only failing trees; and no ten of the young trees can, in this generation, be of half as much interest for an insect-hunter as one of the old worm-eaten fellows cut down to make room for them. All are, however, not yet gone, and there is still in Kensington Gardens at least one tree remaining, a balsam poplar, riddled, like the rock of Gibraltar, with tunnels running in every direction.

The engineers who first drove them were probably the larvæ of Goat Moths, but the lower galleries, to the level at least of an ordinary man's nose, have been long surrendered to Earwigs and Woodlice. The French naturalist, Lyonet, who dissected a Goat Moth Caterpillar, counted as many as 4,061 distinct muscles, of which 228 were in the head. One could have formed a pretty fair idea, without any such detailed information, of the muscular power necessary in the fore-quarters to enable a Grub to work his way through solid timber as easily as if it were so much cheese; and the figures are chiefly interesting, as an instance of the laborious minuteness of the work of many entomologists. "In the human subject," add Messrs Kirby and Spence, from whom the figures are taken at second hand, "only 529 have been counted, so that this minute animal has 3,532 muscles more than the 'lord of creation.'" The smell of a Goat Moth is an instance of the power of association to sweeten the disagreeables of life. Judged on its own merits it is very nasty indeed, but it may be charged with recollections making a whiff as welcome as the scent of mignonette, which now meets us pleasantly at every turn in the park. Such amiable day dreams are, though, if too freely indulged in, liable to a rude awakening. It is not an unheard-of thing for ladies who have been carried back for a long summer afternoon to scenes of childhood, to find, on packing up their sketches, that the source of the fœtid smell in the oak wood, in which the charm had lain, was not, as they had fondly believed it, a wood-witch fungus, but a dead rabbit unnoticed in the fern beside them.

A Caterpillar in a solid tree, asking no better food or bed than the wood, might not unreasonably flatter himself that he was safe from all attack; but if he were to do so he would be very much mistaken.

In the forests of Madagascar, where wood-eating larvæ are found in great numbers, there is a little creature—the Aye-Aye, which seems to have been created, or, if we prefer the phrase, "developed," for the express purpose of keeping them down.

It is in general appearance something between a squirrel and a monkey, and has unusually perfect cutting teeth, eyes, and ears—the last "very large, naked, and directed forwards"; specially fitting it for the kind of life it leads.

To hunt for its food comfortably on the trees it requires to have free use of its hands, and to enable it to do this it has had a clasping thumb given to its hind feet, with which it can hold on to a bough as a monkey does; and, strangest of all, to make its equipment for the life it leads quite perfect, the second finger of the hand, instead of being shaped like all the others, is "slender and long, resembling a piece of bent wire."

"One finger on each hand," writes Professor Owen, who published in the Transactions of the Royal Society in 1863 a very complete anatomical description of the Aye-Aye, "has been ordained to grow in length but not in thickness with the other digits. It remains slender as a probe, and is provided at the end with a small pad, and hook-like claw."

The specimen dissected was kept alive for some little time by Dr. Sandwith, then Colonial Secretary in the Mauritius, and an extremely interesting letter from him is published with Professor Owen's paper. He describes the animal tapping the surface of the worm-eaten boughs put into his cage, "with ears bent forward and nose close to the bark," and poking his slender finger every now and then into the worm holes "as a surgeon would a probe," and when he had made up his mind where to begin, rapidly tearing off the bark, cutting into the wood, and "daintily picking the Grub out of its bed with the slender finger and conveying the luscious morsel to his mouth." A sketch of the hands of the Aye-Aye, taken by permission from one of the pictures of the living animals drawn by Mr. Wolf for the Royal Zoological Society, is given as vignette at the end of the chapter. One of the dainty dishes of the Romans is said to have been made of wood-eating Caterpillars.

Though the Grubs in Kensington Gardens have nothing to fear from Aye-Ayes or human epicures, they have enemies at least as formidable in the very next order of insects which we come to after leaving the Moths and Butterflies.

Of the *Hymenoptera*—one of the most important divisions consists of the "Ichneumons" and other flies like them, which lay their eggs in the bodies of living Caterpillars.

More than one of the Ichneumon flies is armed with a long, sharp, springy "ovi-positor," as it is called, which it either actually bores into trees or pokes through cracks till it finds the soft body of an unsuspecting Caterpillar, into which an egg is slipped, to hatch in good time and eat its unwilling foster-mother. There are numberless varieties of the kind, many of which are believed only to lay their eggs in particular larvæ. The poor old Daddy Longlegs have the questionable honour of an Ichneumon Fly, which seems to confine its attention mainly, if not entirely, to them. "It is impossible," writes Mr. Wood in his little book on "Common British Moths," "to detect a stung Caterpillar till it has ceased feeding, and not always easy to detect it even at that time. Often the Caterpillar changes into a chrysalis without betraying any signs of the mortal injury it has sustained, but when the time arrives for the appearance of the insect, the disappointed collector finds that instead of the Moth the Ichneumon Fly occupies the box." It is satisfactory to know, on the authority of Professor Duncan, that a poetic justice occasionally reaches some, at least, of the murderers. Sometimes the springy ovi-positor, when pressed against the tree, glances from it, and shoots the egg into the last place the mother had intended —her own body—and she flies off to become a living presentment of Milton's image of Sin at Hell's gate, with her children gnawing night and day at her vitals.

Marvels of contrivance meet us at every turn of the page in Natural History.

Knowing the ordinary conditions of feeding-life of almost every kind, one would have supposed it impossible that an animal should live for many days with another creature of comparatively large size living and eating in its tissues, without dying of blood-poisoning, or whatever else may be its equivalent in an insect.

But this danger is avoided by a most strange peculiarity of construction which is found in the larvæ of such parasitical Flies and of Bees. If anyone wishes to know how it is that a beehive is sweet, in spite of the crowds of hungry Grubs crammed into it, or why the juices of a Caterpillar attacked by an Ichneumon are not fatally tainted, he may read a reason in No. XVIII.* of Professor Owen's Lectures on "Invertebrate Anatomy."

In England, if a Caterpillar or Grub escapes the

* See Appendix C.

ordinary diseases to which insects, like all of us, are liable, and is lucky enough to be overlooked by birds, beasts, and other insects, it may, so far as we know, be pretty sure in good time of beginning life again as a Butterfly, or whatever else it may be intended to become.

But in other countries there is another danger to be met. In the insect house in the Zoological Gardens is a case containing what looks like stalks of coarse, irregularly-grown grass, with heavy clumped roots. They are specimens of the larvæ of the New Zealand Swift Moth—the so-called "vegetating caterpillar," which, as the note in the case explains, is liable to the attacks of a fungus (*Sphæria Robertii*) which attaches itself to it and sucks out the juices till the substance of the larva becomes changed to something very like pure vegetable matter. In this state it is eaten as a delicacy by the Maoris. Something of very much the same kind is found also in Ceylon, where the Grub of the Coffee-eating Cockchafer is attacked in the same way by another fungus, which commonly grows an inch or two above the ground and has a yellowish tip. "The Grub," writes a planter who has dug up and examined many of the poor creatures on the lawn in front of his bungalow, " is to be found an inch or two below the surface, always in the position sketched,* the head upwards and body bent as if in suffering."

The cherry tree which Baron Munchausen saw growing out of the stag's head is intelligible. The Baron had himself, for want of shot, fired at it with a charge of cherry stones a few years before

* The sketch referred to is printed as a tail-piece to the chapter on St. Kilda.

St. Hubert's Stag, which carried a cross between its horns, was miraculous. But what is the explanation of hundreds of Cockchafer Grubs being found spiked always between the eyes and never in any other part of the body, by a living sword? Does the seed of the fungus stick to the head and root from the outside, or does the Grub try to swallow it and fail?

If Bacon could have seen a specimen of the fungus blade shooting from the forehead of the Grub he might have left us another chapter of "The Wisdom of the Ancients" on Minerva springing into life from the head of Jove.

Gall Flies—which are vegetables what Ichneumon flies are to insect life, and much more attractive—belong to the same order. They are of many sorts, and lay their eggs on leaves and young shoots of trees, &c., which burst out at the touch into excrescences of many different kinds, some hard and round like woody marbles, others soft and spongy, others brilliantly-coloured hairy tufts, which become the home of the grub when hatched, stored with all the provisions it will require.

The egg-laying wand of every Gall Fly has the magical power of Aladdin's wonderful lamp, and can make a furnished palace spring into existence just when it is wanted.

London vegetation does not seem to be generally attractive to them; but, though they are not by any means so conspicuous in their varieties as in almost any field in the country, Galls of many kinds are to be found in our Parks and Gardens.

It was at one time commonly held that great events might be foretold with certainty by the initiated by cutting open galls and seeing what sort of creatures they contained.

The most familiar and pleasantest of the *Hymenoptera* are Ants and Bees, of neither of which we have any want in London.

Honey Bees, chiefly the Ligurian kind, which are said to be better tempered than the common hardier Black Bees, are kept by several people in the heart of the town, and if fed occasionally in the winter do fairly well. But besides stray members of such hives, we have a fair share of the 250 or more wild sorts known in England, and may watch them at our leisure, of every degree of size and activity, from clumsy "bumblers," which seem never quite sure how to use their wings, and well content if they can blunder home somehow without banging themselves against a tree, to spiritual little Bees which hang for a minute or two at a time, like humming birds, to all appearance motionless over a flower, to vanish with a saucy whisk and appear again the next moment as still as before over another flower.

On a Saturday morning in July, 1885, an assistant in Messrs. Mappin and Webb's shop, while crossing the pavement in Regent Street, found himself suddenly covered from head to waist by a swarm of bees. Fortunately he had presence of mind and kept still until, with the help of sympathetic bystanders, coat and hat were taken off when, as suddenly as it had come, the swarm rose and left him with no hurt more serious than a couple of stings on the neck.

Shakespeare, not very appropriately, puts into the mouth of Henry the Fifth of England, the usurper's son, his description of the Bees which teach

"The art of order to a peopled kingdom."

The story of their merchants, magistrates, soldiers,

"civil citizens," "poor mechanic porters," and "officers of sorts," all looking up to their heaven-appointed sovereign—

> "Who busied with his Majesty surveys
> The singing masons building roofs of gold,"—

and the arguments drawn from them for "the divine right of kings," might have fallen more consistently from the lips of his namesake, Henry the Fifth of France, who chose to live and die in exile at Frohsdorf—" the shore the tide was to reach at the appointed time," which never came—rather than give up his white flag and be a king by compromise.

The Bees have led us too far from London to allow a return to the Ants, which Sir John Lubbock has made his own peculiar property, and we come to the next order, the "*Hemiptera.*" If there were any neat Greek word meaning "hotch-potch," or, better still, "confession of failure," either would be quite as appropriate a name. The order—like the "passeres" among the birds—is the receptacle for the thousands of little creatures which cannot clearly be brought under any other of the accepted orders and contains, as might be expected, what seems to unscientific people a most incongruous lot. The different classes of the order have very little more in common than that, as a rule, they have sucking beaks (Bugs are *Hemiptera*), and do not undergo what is called "complete metamorphosis," that is to say, do not sleep away part of their existence as helpless chrysalis. The name, which of course means "half wings," is, as already explained, intended to convey the idea of an insect having one of its two pairs of wings partly thickened. The order is thus a halfway house between the transparent wings of Bees and Dragon Flies, and the thick upper shields of the Beetles and Grasshoppers; and this is the characteristic of many of its members.

But, as if to break the heart of the beginner, who may have flattered himself that he had at last mastered the first secrets of the entomologist's trade, we find that there are whole families of insects belonging to the order which are wingless. All the known "ocean insects," for instance, are classified as "*Hemiptera.*"

There are several of them—the naturalists of the "Challenger" added to the number—insects not unlike the Gerris, the little slim long-legged Carnivora, which are to be seen racing dryshod over the runs in trout streams, but they have not a wing of any kind among them!

One of the specimens of the order which flourishes much too well in London is the Aphis, the common Green Fly. It is an especial plague of town conservatories, probably because our plants having other difficulties to fight with from which their happier fellows are free in the fresher air of the country, have seldom strength of growth enough to make head against such enemies. The flies seem to come by magic. Two or three appear one day, and a week or two later, if no strong measure is taken, every young shoot is covered thickly.

The explanation is that the Green Fly is among the most rapid, possibly the most rapid, breeder yet known, and, strange to say, is both "oviparous" and "viviparous." That is to say, some of the females give birth to living young; others, born usually later in the season, lay eggs. A single egg laid before the cold sets in survives the winter frosts, and the first warm day in spring or early summer an Aphis is hatched, and almost instantly has a family of a hundred females born alive, each of which, without an unnecessary loss of a day, follows her mother's example. The granddaughters do the same, till ten generations have been born alive. The result, supposing all to live, is $1 \times 100 \times 100$, &c., &c., until in the tenth, not the last generation of the year, the family numbers 1,000,000,000,000,000,000—a quintillion: the figures are Professor Owen's.* Then, and this is strangest of all, comes the eleventh generation. When vegetation is rank in spring and summer, the generations of Green Fly, which have to make hay while the sun shines, cannot spare time for such a slow process as being hatched from eggs. That can wait till there is a use for it, and later in the year the use comes. The Aphis is not hardy enough to survive a sharp winter, and so the eleventh generation of the season—the whole story sounds too like a fairy tale—the eleventh generation is born "oviparous."

When their turn comes to have families, instead of giving birth to living babies they lay eggs, some of which are sure to hatch next spring, and thus save from extinction the race which are the milch cows of the Ants.

We can see our way more clearly again when we leave the quagmire of the *Hemiptera* and come to the *Orthoptera* and *Coleoptera*—Grasshoppers, &c., and Beetles.

One can, when one meets with it, recognise an insect with thick sheaths (κολεός) as upper wings covering a lighter transparent pair, and can understand that there is a difference between those which

* "Invertebrate Anatomy," Lecture XVIII.

have their under wings crossed and those which fold them straight like a fan—straight wings—and is all the better prepared to admit the necessity for separating the two orders, when one is told that there are even more important differences in their earlier stages, the true cross-winged Beetles being, like Butterflies, subject, as a rule to complete the *Orthoptera*, only to partial metamorphosis.

Unless they have all vanished in the last clearance of trees, there are very good specimens to be seen in Kensington Gardens of the curious symmetrical workings of a small tree-destroying Beetle named, from the mischief it can do, "Scolytus destructor." The female forces herself under the rough outer bark of elms and eats her way through the soft tissue between it and the hard wood, dropping her eggs, at regular intervals, to the right and left as she goes.

Each Grub as it is hatched begins working on its own account, and guided by some unaccountable instinct, or perhaps by the position in which the egg is laid, drives a shaft of its own outwards from the centre passage bored by the mother, in a line parallel to that of the brother or sister next to it. The result is a grooved pattern to be seen when, as is sure to follow, the bark comes away, not unlike the clean picked backbone of a sole, excepting that as the Grub grows and needs a wider passage as he travels, the diverging ribs are thickest at the end farthest from the spine.

The "Type-writing Beetle," so named from a fancied likeness of its irregular workings to letters, which does much harm in the pine woods on the Continent, is very much like the "destructor," but devotes itself to firs instead of elms. The ravages of the Type-writing Beetle have at times been as serious as plagues of locusts. In the Hartz forests alone, according to Kirby and Spence, these Beetles—for delivery from which, by-the-bye, there was a special prayer in the old German liturgies—killed in one visitation as many as a million and a half of fir trees.

In another year, it was said, the mines must have been closed, and the country for the time ruined; but happily, just before it was too late, the Beetles took it into their heads to migrate "in swarms like Bees" into other parts where they were probably no more welcome.

But compared with the much smaller insect, the Phylloxera, a comparatively recent importation from America, the Type-writing Beetle is a harmless creature. A French writer, basing his calculations on official statistics, lately estimated the damage done by these tiny creatures in French vineyards only, during the thirteen years from 1875 to 1888, at something like four hundred million pounds sterling.

Before saying good-bye to the poor old sick elms in Kensington Gardens, it may be worth mentioning one other point of melancholy interest in connection with them, though it has nothing to do with insects. Many of them, where the bark has been pulled off by mischievous boys, will be seen to be veined under the bark for several feet above the roots with a curious narrow, flat, dry growth of dark colour, not thicker than paper, but very tough, and clinging so closely as to require a knife to separate it from the wood.

"The growth," writes Mr. Thistleton Dyer, to whom a slip was sent for submission to the learned authorities at Kew, "belongs to an obscure set of organisms known as *Rhizomorpha*. They are not fully-developed vegetable structures, but are really the mycelial portions of large fungi grown under peculiar conditions. In process of time," he adds, "the structure would develop on the exterior of the tree as a large *Polyporus*, a sort of woody fungus. The existence of such a growth under the bark is the tree's death warrant."

When one sees the chain of destruction spreading in every direction through the insect world, as everywhere else, the less one thinks of the "Mystery of Pain"—the subject of one of Canon Kingsley's best sermons in Westminster Abbey—the better for one's peace of mind. But it is some relief to know, even at the cost of loss of faith in the infallibility of an idol, that microscopic anatomy shows that Shakespeare was altogether wrong when he said that—

> " The poor Beetle that we tread on
> In corporal sufferance feels a pang as great
> As when a giant dies."

If it were the case, no Londoner possessing ordinary feelings and a tame Hedge-hog would be able to sleep a wink for thought of the wholesale tortures of which his kitchen would be every night the scene. The services of the Hedge-hog in keeping within bounds the rather disgusting flabby *Orthoptera*—familiarly known to cooks and housemaids as "Black-beetles," but not in the true sense of the word "Beetles" at all—cannot, sentiment notwithstanding, be spoken of too highly.

As a matter of fact, Beetles have extraordinary tenacity of life, and apparently very little sense of pain. A good little boy, who combined with a passion for collecting insects, a tender heart—the two qualities are apt to clash at times, but a boy may be taught to study Nature and make collections without any promiscuous killing—caught and brought home in triumph one day a big Beetle, the catch of his season. It was consigned to a poison bottle, strong enough to have exterminated a family of Butterflies, and taken out in due course dead. But as after a time there was a suspicion of something like galvanic movement in one leg, to make assurance doubly sure, it was put into boiling water, and when all possibility of latent life was past, pinned for the cabinet. Some days afterwards the Beetle was met deliberately walking down the front stairs, carrying with it the pin, which it had drawn from the cork for itself, with as much *sang froid* as if it had been a smart gentleman strolling down St. James's Street, with a gold-headed cane under his arm. Had the lady who first met it had nerve to stay to look, she would probably have seen a malicious grin on the terrible insect's face, as it chuckled at the thought of the mental torture with which it was to repay its captor.

If Beetles have antiquarians among them as well as type-writers, the learned may some day speculate on the origin of the cairn of stones and brick-bats under which the wing-cases, thorax, legs, and all else that was mortal in their champion—the direct descendant of the Beetle-god worshipped in ancient Egypt—slept at last.

Though use has already been made to an extent almost unjustifiable of leave most kindly given by Professor Owen to pilfer from his works, as this is a chapter on Natural History in London, one more passage must be borrowed, because, coming as it does towards the end of the brilliant passage with which he sums up the teaching of his lectures on the "Generation of Insects," it shows how much, if only one knew how to do it, is to be done with London materials.

"Metropolitan duties," he says, "shut out much of the field of Nature; but still she may be found and studied everywhere. I first learned to appreciate the true nature and relations of the nominally various and distinct metamorphosis of insects by watching and pondering over the development of a Cockroach."

There are only two more of the orders named to which our claim as Londoners has to be made good: the "nerved wings" and "screw wings."

The latter are microscopic insects which live during part of their lives as Parasites on Bees. As we have plenty of Bees we have probably also plenty of the insects which live upon them. But as without very good eyes and very good glasses, and time and patience to use them, we are not very likely to find any "screw wings," it is not necessary to say much about them here. Their chief point of interest is that exactly reversing the arrangements of the "two winged" order, their upper instead of their lower pairs of wings are shrivelled up and apparently useless. The last remaining order—the "nerved winged"—is of far more general interest, and contains in the Dragon Flies examples of the most perfect development of powers of flight known, compared with which the wings of a bird are clumsy contrivances.

The bodies of men and other terrestrial animals are comparatively solid masses. Weight is no disadvantage to them. If anything, it is an advantage, as it helps them to force their way through the obstacles which they meet in their comings and goings on the dull earth, and so every available corner of the trunk is packed as tightly as possible.

But with creatures which are to have the power of lifting themselves from the ground the case is different. What they most require is lightness, and so the hollows of the bones, which in men and quadrupeds are used as bottles and casks for holding marrow and other liquids which are wanted to oil the joints and keep the machinery in working order, are in birds turned into dry chambers filled with hot air, lighter, of course, than the cooler air outside, into which the bird has to rise.

But, even with this beautiful contrivance for lightening it, the solid inside frame of bones on which beasts, and birds, and fishes are built is a dead weight to lift; and so in insects, in which the power of flight is carried to a far higher perfection than even in the Swallow or Frigate Bird, different arrangements are made. The heavy skeleton is dispensed with altogether, and instead of it the supports for the body are given by a light, stiff outside skin only; and, instead of having their fluid parts aired as in beasts by little lungs in the middle of the body, they suck in the air through openings, not confined as in men and beasts to mouth and nose, but dotted about the body in many parts, varying in different insects—back, sides, head, and tail—into tubes ("trachiæ," they are called), which run backwards and forwards in every direction through trunk and limbs.

The four powerful, wide-spreading wings of the Dragon Fly, "the Eagle among insects," look as if made of the thinnest goldbeater's skin stretched on wire, stiff at the base and front edges of the wing, spreading out into the finest network. Through almost every thread of the lace, and everywhere else through its body breath passes, so that the Dragon Fly, as one sees it hawking in the fields or over the ponds, is a whiff of living air, imprisoned in a cage of muscle with gripping claws, and a head all eyes and jaw. No wonder with such an outfit, and power to fly backwards and sideways as easily as straightforward, it can kill right and left, and laugh at such poor fliers as the Swallows.

Primitive man having

> "Learnt of the little Nautilus to sail,
> Spread the thin oar, and catch the driving gale,"

it has been reserved, we are now told, for this generation to be taught that all such clumsy contrivances as screws and paddles may be dispensed with if builders will only apply to ships the principle of propulsion which young Dragon Flies have known for thousands of years; from times long before there was a man in the world, for there are traces of Dragon Flies in the rocks of very early dates.

In "The Times" of the 17th October, 1883, is an account of the trial trip of a new German ship driven by hydraulic reaction on the Elbe, near Dresden.

"Noiselessly and without any oscillation," writes the Berlin correspondent, "did the large vessel—large as compared with the steam-craft plying on that part of the river—after the simple turning of a lever by the captain on the bridge, commence its trial-trip, stemming the current and keeping an even course under the picturesque right bank of the river. The only noise audible was that of the rushing of the water from the tubes, fixed a little above the level of the river, and nearly amidships, on both sides of the vessel. Another turn of the lever, and the action was reversed. The vessel comes to a dead stop in less than her own length."

The larvæ of the Dragon Flies shoot themselves along in just the same way by squirting through tubes. The principle—which had already been applied in the British Navy in the Waterwitch—is exactly that of the German invention, and is destined, according to Admiral von Henck, "to modify or even supersede the present ship engines." The only difference is a trifling one in the application. The valves, instead of being "amidship," and "above the water" as in the vessel, are in the insect placed at the stern and used under water.

Dragon Flies of several kinds are occasionally, though not very often, to be seen in London, but we have plenty of other representatives of the "nerve winged" order. A very pretty one, with antennæ much longer than itself, is to be seen in quantities any warm day in August or September, running about on the flat wooden railing of the bridge across the water in St. James's Park. It is what fly fishermen know as the "Cinnamon," one of the Cadis Flies—the family which, in the grub state, live in the water and build themselves quaintly-ornamented houses, which they carry about with them. The nerves of the wings of the Cinnamon Fly are almost hidden by brown feathers, but it is still one of the "Neuroptera." There are plenty of the "Ephemera," too, of which the "May Fly" is one. May Flies in the perfect state have no visible mouth, and no one yet has been able to discover that they take food or nourishment of any kind, unless it is sucked in with the air through their breathing-tubes.

One of the many things which Naturalists have yet

to learn is the explanation of the appearance of thousands of the same fly at almost the same moment. Not a May Fly is to be seen for months, and then, no one can tell why, all at once the trout are leaping at them in every direction.

This is even more remarkable on some of the German rivers, where on a warm evening swarms of a large, light-coloured Ephemera come out suddenly, till they look like a thick mist on the water. Straw fires are lit on the banks, and next morning baskets full of yellow-bodied flies, with white wings singed, are swept up for the poultry. Why is it, too, that probably within five minutes of one another all the Bats in a neighbourhood wake and come out?

The habit is to be noticed in London as well as anywhere. There is at least one tree in Kensington Gardens—an old hollow oak between the refreshment room and the gardener's cottage—which is the home of a considerable colony of Bats. A note was made of the exact hour at which the long silent procession left the hole one evening in August. The next day, within four minutes of the same time—the time was carefully taken—seventeen Bats crawled up, and with the same regular intervals took headers into the dusk, to appear again as if they had started from another quarter altogether, careering about over the tops of the trees, doing the best they could to prevent too great an increase of humbler London Night Fliers.

The part which insects and kindred small creatures of the kind have played and still play in the creation and support of the world as we now live in it has been a theme for many pleasant writers. Whole island groups, described by sailors who have seen them as heavens on earth, owe their beginnings to the work of the little Coral Polyps, who are responsible too, in all probability, for the present varied coast-lines of large tracts of continents. There are families of plants, some of rare beauty, which we are told would disappear entirely but for the busy Bees which, forcing themselves in and out in search of the honey baits placed there to tempt them, carry the fertilising dust to the seed-vessels, which must otherwise have dropped useless.

But the chapters which are to tell of the influence of insects on social and political history have yet to be written. With justice done to the subject they should be very interesting. Flies, Lice, and Locusts did, as we all know, a good deal to help Moses in his struggle for Hebrew independence; and Wasps and Hornets, when Jordan had been crossed, fought on the side of Israel, "forerunners of the host." Pyrrhus, at the height of his successes, raised a siege because he could not stand the Mosquitos, which made his camp unbearable.

> "If 'ifs' and 'ans' were pots and pans,
> There'd be no work for tinker's hands."

A great many things might have been told differently in Roman history—though it is not easy just now to say exactly how—if Pyrrhus and his Greeks had added to their more brilliant qualities the patient obstinacy which made Sir Henry Lawrence's band hold out in Lucknow till Havelock came, in spite even of "infinite torture of flies."

Difference of race may be at the bottom of much that is to be seen north of the Tweed and across St. George's Channel; but making all allowance for this, the legacies of conquest will not be lost sight of altogether. And Scotland was once, as completely as Ireland ever was, a conquered country, and with every prospect of remaining so, until a Spider came to the rescue, and by cheering the Bruce, as he lay with broken spirits—six times beaten—to make a last attempt which was to succeed, drove out the English and delayed the Union for three hundred years, until all reasonable excuse for soreness was forgotten.

To look further still afield and take a wider view. It is a simple matter of history that the danger faced on the field of Tours did more than anything else to startle Europe, which had lain for a long time half stunned and stupefied by the fall of Rome, into life again. And surely not even the modern sceptics, who throw doubt on the Dun Cow, and try to explain away Lady Godiva's ride through Coventry, would wish us to give up the sure belief that we should never have heard of the Saracens but for that other Spider which stood ready at Allah's command to throw his web, at a critical moment, over the hole in the cave in which Mahomet crouched for his life on the great day of the Hegira—the day which is to every Mussulman, good, bad, or indifferent (some five or six times as many, according to Mr. Bosworth Smith's estimate, as the entire population of the British Isles), what the year of our Lord is to Christendom. Then there would have been no Mahomedan invasion to give Charles Martel the opportunity to put his mark on his century, and clear the way for Charlemagne to found the second Western Empire.

When Mahomet's Spider shook himself that morning and thought he would try to catch a fly for his breakfast, he turned the entire course of mediæval and modern history, smashed the great Persian empire, burned the library of Alexandria, and, among other things—but we are following him on to dangerous ground, and had better stop before quite over head and ears—laid the first stone of the fabric of the temporal power of the Popes. The chain of cause and effect all through is unbroken, and is spun with the Spider's thread.

It is an aid to faith to know that the little creatures which worked such miracles are not mere insects, but form, with Scorpions and some other smaller things (including Cheese Mites), a distinct class, which is by common consent placed above the true insects.

The reasons for the higher place given to these bloodthirsty little creatures are that, both in head and breathing contrivances, they are more fully developed, and that, unlike most insects which are born in a shape quite unlike their fathers and mothers, and only reach the perfect form after a series of metamorphosis, young Spiders leave the egg scarcely less like their parents than a human baby is to a grown man or woman.

The female Spider is almost always larger and stronger than her husband, and is prepared at any moment to dine upon him if he comes too rashly near her.

The etiquette of the Court of the Medes and Persians of Esther's day, which made it death to the Queen who approached the King without a summons unless he held out his golden sceptre, is exactly reversed by the Spiders.

The terrible lady sits in state in the middle of her web, and, if willing to receive her lord, lets him know it by raising the front pair of her eight legs in token of acceptance. He, unless very young and foolish, knowing her temper well enough and what she is capable of, when in a passion, first introduces himself by cautiously shaking the web at a safe distance. If rash or inexperienced enough to go within reach of her jaws without receiving the sign of acceptance, he is not likely to have a chance of repeating the indiscretion.

The insect world is enchanted land, and in it we are apt to wander on as forgetful of time and space as was Rip Van Winkle in the Sleepy Hollow. But one must stop somewhere, and having strayed over the border to the Spiders it will be wise to rest there without wandering farther.

The subject of one of Artemus Ward's lectures as published in his advertisements was "The Babes in the Wood." After an hour's talk on every other conceivable subject he apologised for digression, and said that he had every reason to believe that the babes in the wood were "very nice young people," but he did not think he had anything else to say about them. It is difficult to help feeling a little uncomfortably suspicious of being guilty of something very like a feeble imitation of his style on coming to the end of a chapter on "London Insects," which gives no list and mentions scarcely a dozen of them. But it is a charm of Natural History in all its branches that nothing that is of human interest can be altogether strange to it. The students of "God's great second volume," even in the lowest forms, are franked to wander where they will through playgrounds to which the Yellowstone Park is nothing, and, grey hairs notwitstanding, while there are schoolboys still.

One of the greatest disadvantages of life in a great town is its artificialness. Man and his works are everywhere, and Nature is lost sight of in the dust and din; and if these sketches should by any lucky chance be the means of calling a town reader's attention for the first time to the wild birds and insects of London, and showing him that we have really some to take an interest in, they will have awakened a new power of enjoyment, and done more than he may at first suppose to add to his happiness.

THE END.

APPENDIX A.

The following list of wild birds noticed at different times in London is based upon one drawn up by Edward Hamilton, Esq., M.D., F.L.S., published in the "Zoologist" of July, 1879.

The birds marked with one asterisk may be described as "*casual;*" those marked with two asterisks as "*rare casual*" visitors. The others are more or less constant residents during some part of the year.

RAPTORES.

**Peregrine Falcon	*Falco peregrinus.*
**Kite†	*Milvus regalis.*
*Kestrel	*Falco tinnunculus.*
*Sparrowhawk	*Accipiter fringillarius.*
*Barn Owl	*Strix flammea.*

PASSERES.

**Great Grey Shrike	*Lanius excubitor.*
Spotted Flycatcher	*Muscicapa grisola.*
Missel Thrush	*Turdus viscivorus.*
*Fieldfare	*Turdus pilaris.*
*Redwing	*Turdus iliacus.*
Song Thrush	*Turdus musicus.*
Blackbird	*Turdus merula.*
**Ring Ouzel‡	*Turdus torquatus.*
Hedgesparrow	*Accentor modularis.*

† Once built constantly in the trees in London.
‡ These birds are occasionally seen on migration.

Appendix.

PASSERES—(continued).

Redbreast	*Erythaca rubecula.*
Nightingale	*Sylvia luscinia.*
Redstart	*Ruticilla phœnicurus*
*Wheatear	*Saxicola œnanthe.*
*Reed Warbler	*Salicaria arundinacea*
*Sedge Warbler	*Salicaria phragmitis.*
*Greater Whitethroat	*Sylvia cineria.*
*Lesser Whitethroat	*Sylvia sylviella.*
*Garden Warbler	*Sylvia hortensis.*
Blackcap	*Sylvia atricapilla.*
Wood Wren	*Sylvia sylvicola.*
Willow Wren	*Sylvia trochillus.*
Chiffchaff	*Sylvia rufa.*
*Golden-crested Wren	*Regulus cristatus.*
*Tree Creeper	*Certhia familiaris.*
*Nuthatch	*Sitta Europœa.*
Great Titmouse	*Parus major.*
Blue Titmouse	*Parus Cœruleus.*
Cole Titmouse	*Parus ater.*
*Marsh Titmouse	*Parus palustris.*
Longtailed Titmouse	*Parus caudatus.*
Pied Wagtail	*Motacilla yarrellii.*
*Grey Wagtail	*Motacilla boarula.*
*Ray's Wagtail	*Motacilla flava.*
*Tree Pipit	*Anthus arboreus.*
Meadow Pipit	*Anthus pratensis.*
*Sky Lark	*Alanda arvensis.*
*Common Bunting	*Emberiza miliaria.*
Chaffinch	*Fringilla cœlebs.*
House Sparrow	*Passer domesticus.*
*Hawfinch	*Coccothraustes vulgaris.*
*Yellowhammer	*Emberiza citrinella.*
Wren	*Troglodytes vulgaris.*
**Golden Crested Wren	*Regulus cristatus.*
Greenfinch	*Coccothraustes chloris.*
*Goldfinch	*Carduelis elegans.*
**Mountain Finch	*Fringilla Montifringilla.*
*Siskin	*Carduelis spinus.*
Lesser Redpoll	*Linaria minor.*
Linnet	*Linaria canabina.*
*Bullfinch	*Pyrrhula vulgaris.*
Starling	*Sturnus vulgaris.*

Appendix.

PASSERES—(continued).

**Raven	*Corvus corax.*
*Carrion Crow	*Corvus corone.*
*Hooded Crow	*Corvus cornix.*
Rook	*Corvus frugilegus.*
Jackdaw	*Corvus monedula.*
Magpie	*Pica caudata.*
*Kingfisher	*Alcedo ispida.*
Swallow	*Hirundo rustica.*
Martin	*Hirundo urbica.*
Sand Martin	*Hirundo riparia.*
Swift	*Cypselus apus.*
**Nightjar†	*Caprimulgus Europæus.*

SCANSORES.

Green Woodpecker	*Picus viridis.*
**Great Spotted Woodpecker	*Picus major.*
**Lesser Spotted Woodpecker	*Picus minor.*
**Wryneck	*Yunx torquilla.*
*Cuckoo	*Cuculus canorus.*

GALLINACES.

Ring Dove	*Columba palumbus.*
Rock Dove	*Columba livia.*

GRALLÆ.

**Woodcock	*Scolopax rusticola.*
*Common Snipe	*Scolopax gallinago.*
**Brown Snipe	*Scolopax griseus.*
*Dunlin	*Tringa variabilis.*
Moorhen	*Gallinulina chloropus.*
Coot	*Fulica atra.*
**Grey Phalarope	*Phalaropus lobatus.*
Heron‡	*Ardea cineria.*

† These birds are occasionally seen on migration.
‡ Occasionally to be seen flying over London.

PALMIPEDES.

Mallard	*Anas boschas.*
*Teal	*Anas creca.*
Widgeon	*Anas penelope.*
Little Grebe	*Podiceps minor.*
*Common Tern	*Sterna hirundo.*
*Kittiwake	*Larus tridactylus.*
*Common Gull	*Larus canus.*
Herring Gull	*Larus argentatus.*
Black-headed Gull	*Larus ridibundus.*
*Cormorant	*Phalacrocorax carbo.*
*Guillemot	*Uria troile.*

APPENDIX B.

The following instance of the perseverance with which a bird will, at times, cling to the spot selected for a nest in the face of what might be supposed insuperable difficulties is given as told by a careful and patient observer of nature, who has himself succeeded in taming free swallows, and inducing them to feed their nestlings on his lap.

One is tempted on reading the story to wonder that the sternest of Lady Abbesses,—

> "Though vain of her religious sway,
> Loving to see her nuns obey,"

could have seen the bird complete the round of the nitches of the marble grave, and begin a second round, without relenting.

"I have a sister who is a nun in the ——— Abbey, at ———. When she was at home she took part in much of the Natural History work my brothers and I were engaged in, and she certainly has a good knowledge of our English birds, and is a careful observer. I mention these facts to show you that my witness is one on whom I can thoroughly rely.

"On Thursday, July 9th, last year (1891), a Missel Thrush built in the Nuns' graveyard. There is a marble grave there with a pent house built over it to save it from the weather. The rafters resting on the "wall-plate" make twelve openings under the eaves, six on each side of the building.

"In the first of these openings, on the left, the nest was

built. The Lady Abbess told my sister to take it out, as she feared for the future of the grave stone when the young birds would be hatched. It was removed on the 12th, and by the afternoon of the next day another nest was nearly finished in the next compartment between the rafters. This was promptly removed, but on the 14th the third nest was more than half built in the next division of the roof.

"The bird built in this way all down one side of the roof and then down the openings of the other.

"The nests were removed every second or third day, but it had no effect in stopping it.

"When it had been down the second side of the roof, it began again where it had made the first nest and worked steadily down the openings again, almost always in order, and then round to the other side once more.

"As the nests were taken out they were stood on the ground at the foot of a wall that bounds the graveyard. On this wall there are marks showing where the line of nests began and ended—a distance of over ten feet, for this extraordinary bird built over thirty nests.

"My sister herself took eight and twenty, and another nun, who also looked after this part of the grounds, says she took out 'a great number,' but she did not keep account.

"The bird built even while closely watched, and I am told 'flew about screaming as if it were mad.'

"It would snatch up green weeds the nuns had just pulled up as they threw them down, for it got hard up for building material by about the middle of August. It was only stopped at last by my sister pushing holly into the holes, and so preventing any further building operations. Strange as this may sound, I have every reason from knowing many of the witnesses to believe that it is most perfectly true."

APPENDIX C.

"It is remarkable that the larva of the Bee and of the parasitic Hymenoptera have no anal outlet; no fæces are passed until the larva has acquired full growth, and has ceased to feed, preparatory to the pupa-state : thus the fluids of insects infested by the parasitic larvæ are not contaminated by the excrements of their parasites; and the Bee-cells are kept sweet and clean during the active life of the larva."—"Lectures on the Comparative Anatomy and Physiology of Invertebrate Animals." By RICHARD OWEN, F.R.S. (Second Edition.)

www.ingramcontent.com/pod-product-compliance
Lightning Source LLC
Chambersburg PA
CBHW020249170426
43202CB00008B/297